A NATURALIST'S GUIDE TO THE

BIRDS OF COSTA RICA

Steve Bird

JOHN BEAUFOY PU

Reprinted in 2022

This edition first published in the United Kingdom in 2019 by John Beaufoy Publishing Ltd
11 Blenheim Court, 316 Woodstock Road, Oxford OX2 7NS, England
www.johnbeaufoy.com

10 9 8 7 6 5 4 3 2

Photo Credits

Front cover: *main image* Resplendant Quetzal; *bottom left* Yellow-throated Toucan; *bottom centre* Silver-throated Tanager; *bottom right* Coppery-headed Emerald (all Steve Bird). **Back cover:** Flame-coloured Tanager (Steve Bird). **Title page**: Emerald Tanager (Steve Bird). **Contents page**: Keel-billed Toucan (Steve Bird).

Main descriptions: All photos by **Steve Bird** except as detailed below. Photos are denoted by a page number followed by t (top) or b (bottom).

John Ashworth 119b; **Jim Carr** 43t; **Frank W Mantlik** 18b, 19t, 27b, 30b, 34t, 39b, 46t, 48t, 48b, 56b, 74t, 89t, 93b, 109t, 118t, 131t, 135b, 147b, 152b, 153t; **Gina Nichol** 9, 10, 13, 90t, 107t; **John Schwarz** 46b, 49t, 50t, 51t, 87b, 107b, 122b, 135t.

ISBN 978-1-912081-02-8

Edited by Krystyna Mayer
Designed by Gulmohur Press, New Delhi

Printed and bound in Malaysia by Times Offset (M) Sdn. Bhd.

·Contents·

INTRODUCTION

Costa Rica is located in Central America, between Nicaragua to the north and Panama to the south, with the Pacific Ocean to the west and the Caribbean Sea to the east. It measures only 464km from north to south, and 274km at its widest from east to west. Despite its small size, its varied habitats are all home for wildlife. In fact, Costa Rica holds nearly one-tenth of the world's bird species, with 903 recorded to date. With a current population of 4.8 million people, most of whom are concentrated in the capital city, San José, and larger towns and cities such as Alajuela, Cartago and Limón, this leaves more than 25 per cent of the country with many areas of protected rainforest, national parks, private reserves and vital habitats. In fact, Costa Rica is recognized as a world leader in conservation policies on environmental protection and sustainable ecotourism, making it a role model for many other countries.

Costa Rica is the perfect destination for birdwatchers, with excellent infrastructure, reasonably short distances to drive between sites, some excellent lodges and a wealth of fantastic wildlife. Additionally, belonging to Costa Rica and some 550km offshore, Cocos Island is a verdant paradise with lush rainforest and a rugged coastline where some 87 bird species including three endemics (see p. 8) have been recorded.

This book introduces some of the most readily encountered bird species in Costa Rica. It features a large selection of both resident and migratory species, including most that you could expect to find with some certainty if visiting the right habitat at the right time of year. There are, of course, a great deal more species in Costa Rica than can be included here, but many of these are scarce migrants, or notoriously difficult or unpredictable to find. For the casual birdwatcher and nature lover, the book is a good introduction to Costa Rica's most regularly seen avifauna, and a handy guide to the many birdwatching opportunities and places to see such wonderful birds. Each species account includes details of some of the sites, including specific lodges, where you stand a good chance of seeing the bird. It should be noted that this is just a sample of a few places where the bird may be found, and the information is by no means extensive.

GEOGRAPHY

Over the centuries the geography of Costa Rica has been influenced by countless earthquakes, floods and volcanic eruptions, all of which have helped shape its present-day landscape. Dominating this dramatic landscape is a central mountain range that runs the entire length of the country from the north-west to the south-east. The mountain range greatly influences the climate, and contributes to the huge variety of habitats and vegetation that makes Costa Rica such a rich haven for wildlife.

The five towering mountain ranges consist of the Central Mountain Range (Cordillera Central), Talamanca Mountain Range (Cordillera de Talamanca), Tilarán Mountain Range (Cordillera de Tilarán), Guanacaste Mountain Range (Cordillera de Guanacaste) and Escazú Hills (Cerros de Escazú). The highest point in the country is Mt Chirripó at 3,820m, located in Chirripó National Park within Cordillera de Talamanca. Additional to this is Irazú Volcano, located in the Cordillera Central and the tallest volcano in the country, reaching upwards of 3,432m. It is one of the five active volcanos that include Poás Volcano, Arenal

Volcano, Turrialba Volcano and Rincón de la Vieja Volcano. Costa Rica has more than 60 volcanos due to the shifting of two tectonic plates, the Caribbean Plate and the Cocos Plate, which have also caused many earthquakes over the years.

Apart from the mountains, most of the geography is made up of four main geographical areas: the Tropical Lowlands of the Caribbean and Pacific coasts, Northern Central Plains, Central Valley and Northwest Peninsula (see map, inside back cover). As mentioned opposite, there is also Cocos Island in the Pacific Ocean, a designated national park. For convenience, in this book the country has been split into areas that relate to habitats within the species information; these include the Central Valley, the North and South Pacific Slopes, the North and South Caribbean Slopes, the Highlands, the Osa Peninsula and the Nicoya Peninsula.

CLIMATE

There are basically two seasons in Costa Rica: the dry season or summer (December–April), and the wet season or winter (May–November). The climate is considered tropical year round, but because of the varying terrain and habitats there are also many microclimates, making it easy to drive from hot, dry areas to cold, wet ones in a few hours. Both the rainfall and humidity are highest on the Caribbean Slope, with an annual rainfall of nearly 50cm, although typical wet-season rain occurs mostly in the afternoons. The temperature can vary from around 30 °C in the coastal lowlands to 10 °C on the tops of the highest mountains. Due to cloud cover on the Pacific coast, the temperatures are slightly higher than in other parts of the Caribbean. Surprisingly, the best time to see birds is between November and late March, when activity is at its greatest. Light rain is by far the most productive environment in which to see birds, because in such conditions they become very active, and display, sing and show themselves off. In warm or hot temperatures the birds become less active, and are consequently often quiet and difficult to find. If it looks as though the day will be hot or warm, go out as early as possible before the temperature rises and bird activity ceases.

HABITAT TYPES

Due to Costa Rica's geographical position between the continents of North America and South America, its Neotropical climate and its wide variety of habitats, it is home to nearly 4 per cent of the total estimated bird species in the world. For such a small country this is truly astounding.

LOWLAND WET FOREST

This includes the classic tropical rainforest type, with both deciduous and evergreen emergent trees reaching heights in excess of 45m. Huge tree buttresses are common and many trees have broad, round canopies. The shady understorey may consist of large-leaved plants and palms, climbers, lianas and strangler vines, herbs, shrubs and small trees, often with a bed of leaf litter and rotten fallen tree trunks. Many of the large trees are covered with epiphytes and bromeliads, which are home to countless insects. These forests are interspersed with small streams.

This is a favoured habitat for a wide range of birds such as antbirds, tinamous, manakins, puffbirds, wrens and woodcreepers. National parks that fall within the Caribbean lowlands include La Selva Biological Reserve, Caño Negro National Wildlife Refuge, Tortuguero National Park and Cahuita National Park, while southern Pacific lowland parks include Carara National Park, Corcovado National Park and Manuel Antonio National Park.

LOWLAND DRY FOREST

This is the most common habitat of the northern Pacific lowlands (Guanacaste province). It consists mostly of low deciduous trees and shrubs, usually no higher than 7–9m and with scattered taller, 25m trees among them. The understorey is often dense with vine tangles and varieties of spiny thorn scrub and cacti. This habitat is favoured by fewer species than lowland wet forest, but it includes several trogons, parrots, Great Curassow, Thicket Tinamou, gnatcatchers, wrens, euphonias and hummingbirds. The two parks that dominate this area are Palo Verde National Park and Santa Rosa National Park.

HIGHLANDS AND PACIFIC AND CARIBBEAN SLOPES

Costa Rica is dominated by highlands that consist of wet montane forest, cloud forest, paramo, elfin moss forest and montane oak. It is the highlands and Caribbean and Pacific Slopes that hold the greatest diversity of life. Nearly 50 per cent of all the species are endemic or near endemic, sharing this habitat with western Panama. Life can be harsh especially the higher you go, and as you work your way higher, so the species seem to change with every 100m gained. There is often mist and fog and sometimes endless rain, but birds, insects, plants and other life thrive here, and it is easy to see hundreds of birds and other animals during a visit of a few weeks. The habitats are varied, with many of the forests literally dripping with epiphytes, mosses, bromeliads and orchids, while the stunted elfin forests and low scrub of the paramo offer a totally different scene.

Here in the highlands and on the middle- to high-elevation slopes, birds join together into mixed feeding flocks. To witness one of these flocks (with 30–40 species) pass by you in quick succession can be one of the most thrilling experiences in the natural world. It is here that you will come across numerous tanagers, warblers, flycatchers, wrens, woodcreepers, hummingbirds, thrushes, vireos and guans, and the holy grail of the birds and possibly the most beautiful bird in the world – the Resplendent Quetzal. Among the parks that host this type of habitat are Arenal Volcano National Park, Braulio Carrillo National Park, Irazú Volcano National Park, La Amistad National Park, Monteverde Cloud Forest Reserve, Poás Volcano National Park, Rincón de La Vieja National Park, Wilson Botanical Garden, Chirripó National Park and Tapantí National Park.

WETLANDS AND COASTAL HABITAT

One of Costa Rica's best wetland areas is the flood basin of Palo Verde. Here an extensive area floods during the wet season and gradually dries out during the dry season. The 'Palo Verde' tree is predominant here, and the vast open basin is covered with water hyacinths, sedges and cattails. This excellent area can concentrate thousands of water birds such as whistling ducks, gallinules, herons, ibis, egrets, Snail Kites, Limpkin and jacanas, and it is the best place in which to look for the huge Jabiru. Another superb wetland area is Tortuguero National Park, most of which is an alluvial floodplain consisting of a network of blackwater canals and creeks, Raphia Palm swamps and rainforest. Boat trips within the park offer chances to see a wide variety of wildlife, including many herons, kingfishers, hawks and parrots. The beaches at Tortuguero (meaning turtle) are a globally important nesting site of the endangered Green Turtle, as well as Leatherback, Hawksbill and Loggerhead Turtles.

Boat trips on several rivers, such as the Rio Sarapiqui and Rio Caño Negro, are available. Depending on the water level and season, these can provide great opportunities to see a variety of often difficult-to-view birds such as kingfishers, wood rails, Green Ibis, Mangrove and Southern Rough-winged Swallows, and (with careful searching) both the Sungrebe and Sunbittern, as well as huge Green Iguanas. The boatmen are very experienced and offer relaxed and enjoyable wildlife viewing.

Another superb wetland area and a must for any visitor is the Rio Tárcoles. Here you can take a morning or evening boat trip that will allow you to view many bird species up close along this tidal river, river mouth and nearby mangroves. The experienced boatmen are mostly good birders and can point out many of the species, such as the Mangrove Hummingbird, Double-striped Thick-knee, American Pygmy Kingfisher, Boat-billed Heron and hundreds of other water-loving birds. Areas of coastal mangrove mostly along the Pacific coast are characterized by Red and Pacific Mangroves, and are a haven for a wide variety of birds. The Boat-billed Heron, Common Black Hawk, Crane Hawk, Mangrove Hummingbird, Mangrove Vireo and Panama Flycatcher, and even the rarely seen Rufous-necked Wood Rail, all exist in this dense habitat.

Taxonomy

The information in this book is based on the sequence, nomenclature and taxonomy of the World Bird List version 9.1, published by the International Ornithological Congress (IOC), which is considered to be the most authoritative treatment on bird taxonomy relative to Costa Rica. The list captures most of the recent taxonomic changes, which have resulted in revisions of many English and Latin bird names, and classification at higher taxonomic level. Note the use of the English spelling 'grey' instead of the American 'gray', in keeping with the series style of the Naturalist's Guides.

The Bird Species

With currently 903 bird species now recorded in this relatively small country, there is certainly more than enough to keep even the most avid birder busy. More than 630 of these birds are considered resident, and the rest are mostly migrants from North America. Of particular interest are the endemic and near endemic species – birds found either only in Costa Rica or only within Costa Rica and western Panama.

ENDEMIC AND NEAR ENDEMIC BIRDS

There are eight true endemics (bird species occurring only in Costa Rica). Of these, five are found on the mainland: the Mangrove Hummingbird, Coppery-headed Emerald, Black-cheeked Ant Tanager, Dark Pewee and Grey-tailed Mountaingem. Three are found on the tiny Cocos Island: the Cocos Cuckoo, Cocos Flycatcher and Cocos Finch. There is a group of [illegible] additional species that are considered near endemics as they only occur in Costa Rica and Panama.

INTRODUCED AND INVASIVE BIRDS

Some of Costa Rica's birds are not native and have been introduced from other parts of the world. There are just four introduced species, and these are the Rock Dove, Eurasian Collared Dove, House Sparrow and Tricolored Munia.

MIGRANT BIRDS

Many long-distance migrants spend half the year away from their breeding grounds in North America to reside in Central and South America. A few species, such as the Swallow-tailed Kite, Plumbeous Kite, Yellow-green Vireo and Piratic Flycatcher, migrate from South America early in the year to set up territories and breed in Costa Rica and further north. Most migrants, however, travel from North America, with the majority being warblers, thrushes, vireos, hawks and shore birds. Birds heading south generally tend to arrive along the Caribbean coast by early August, with numbers dramatically increasing in September and October. Not all stay and many of these continue on to South America. By mid-March they start to head north again, with most migrants being visible during April into early May. Some of the most impressive migrations involve raptors such as Swainson's Hawks, Broad-winged Hawks and Turkey Vultures; literally thousands of these birds can be seen drifting north in favourable weather conditions.

Identifying Birds in the Field

By using your ears and trying to learn some of the forest bird sounds either through persistence or by listening to some of the many superb bird songs available on CDs and apps, you will soon discover that the seemingly quiet forest is actually full of bird life. Carry an identification book with you, as well as a good pair of binoculars and notepad, and if you can, take a guided tour with a local expert or even book yourself onto one of the many organized birdwatching tours that can be found through the Web or local bird clubs.

Where to Watch Birds in Costa Rica

The following 'hotspots' represent a selection of the best-known sites for birdwatchers, where a good range of birds and other wildlife can regularly be seen. Many of these areas, and more specifically some of the lodgings within or close to them, are referred to in the sites section in each species account.

ARENAL VOLCANO NATIONAL PARK

This 2,900ha area is within the Arenal Tilarán Conservation Area, which itself protects eight of Costa Rica's 12 life zones and 16 protected reserves in the region between the Guanacaste and Tilarán mountain ranges, and including Lake Arenal, and is dominated by the Arenal Volcano, Costa Rica's most active volcano, regularly emitting columns of ash from the crater. For the birdwatcher this is a nice distraction, but of more interest is the national park nestled below the volcano. Here the combination of moist wet forests, streams and nearby Lake Arenal provide habitat for many different birds, mammals, reptiles, amphibians and plants.

For those not afraid of heights, the nearby Arenal Hanging Bridges offer the chance to walk a 3km trail through forest habitat, crossing 15 suspension bridges, the tallest of which is 60m high and the longest 98m long. Here you can often view birds of the canopy eye to eye. Secondary growth and roadside birding near Arenal Volcano NP can be very productive, especially along the roads leading towards Arenal Observatory Lodge.

Arenal Volcano

BRAULIO CARRILLO NATIONAL PARK

This park is just a short distance north of the city of San José, which makes it easy for a day visit. Situated in the rugged Central Highlands, it consists mostly of middle- to high-elevation rainforest, cloud forest, and rivers and streams. Beside the Quebrada Gonzales Ranger Station there is a short but rather steep (in places), circular trail. This takes you through prime rainforest habitat that can seem either devoid of life or completely overwhelming, depending on the weather, timing and luck.

What makes regular trips here worthwhile is the possibility of coming across a mixed bird species flock that suddenly appears with 40 or more species noisily dashing through the forest. There are several really rare and uncommon species here, and quietly walking the trail may reward you with sightings of some of these special birds. Created in 1978, the park comprises five life zones and some 6,000 plant species, 500 bird species and nearly 150 mammal species. Just a little further along the highway you will find Braulio Carrillo Aerial Tram, which offers a wonderful way to experience the rainforest. The aerial tram, which covers 4km in about an hour and 20 minutes, takes you smoothly through the tree canopy, where you can see many birds and mammals. Note that there are entrance fees to this park.

CARARA NATIONAL PARK

This wonderful national park was established in 1978, and being just 48km west of San José it is a must for any visiting birdwatcher and nature lover. The park protects the river basin of the Rio Grande de Tárcoles, near Orotina, Puntarenas, on the Pacific coast, and mainly consists of primary and secondary rainforests, lakes, rivers and waterfalls.

There are two trails that offer excellent birdwatching opportunities. The first is about a kilometre from the Rio Tárcoles bridge, where you may want to make a quick stop to look at the huge American Crocodiles that roost on the river banks below the bridge. The second is a loop trail a little further on and accessible from the Quebrada Bonita Ranger Station, where you will need to purchase a day ticket. By far the best time to visit either trail is early morning, as it can get busy with tourists and school groups throughout the day. Bird activity is best in the morning and you need to take your time to look for birds. Most people walk too fast and miss the wealth of wildlife that exists here, including numerous reptiles, amphibians, insects, plants and, of course, birds and mammals.

During the day, small streams can be excellent for seeing small passerines such as manakins and warblers bathing, shady areas are good for quail-doves and antpittas, while flowering shrubs can attract a variety of hummingbirds. This is also the best place in the country to see the superb Scarlet Macaw.

RIO TÁRCOLES

This river borders the Carara National Park and can be used to combine an afternoon boat trip with the aforementioned early-morning visit to Carara. The river is 111km long, but we are only interested in the area from the estuary mouth to the river bridge just a few of kilometres inland. This is unfortunately the most polluted river in Costa Rica, but surprisingly it holds an incredible diversity of birds. Several companies offer wildlife boat trips in the morning or afternoon, and these provide the opportunity to get close to many species of heron, ibis, kingfisher and shore birds, and some of the famously huge American Crocodiles.

Of particular interest to birdwatchers will be a trip into the tributaries that go deep into the mangroves. Here the American Pygmy Kingfisher, Mangrove Hummingbird, Boat-billed Heron and a host of other birds can all be seen up close. Probably the best-known boat trip is operated by Luis Campos of Mangrove Birding Tours. They know the birds and cater for many bird groups and individuals, offering a very professional service; in the last few years some of the other boat operators have also got really good at knowing the birds and can provide an excellent experience.

The estuary often holds a roost of terns and gulls, alongside Brown Pelicans, many shorebirds and the ever-present Magnificent Frigatebirds. Further upriver towards the bridge is good for the Double-striped Thick-knee, Collared Plover and Southern Lapwing, but note that there are boat trips to this section of the river that go to put on a show by feeding the crocodiles, something you may or may not like to see. The edges of the river also provide habitat

Rio Tárcoles mangrove

for many other species apart from the water birds, and Ospreys, Common Black Hawks, Mangrove Swallows, Yellow-headed Caracaras, and a variety of parrots and flycatchers can all be seen.

CAÑO NEGRO NATIONAL PARK

Located close to the Nicaraguan border, this vast wetland is centred on Lake Caño Negro, a 800ha freshwater shallow lagoon fed by the Frío River during the rainy season. Established in 1984 to protect an area of particular importance to millions of migratory birds from North and South America, the park is mostly accessible by boat or canoe.

Organized boat trips can be arranged if staying at Caño Negro Lodge or Hotel de Campo, and these trips allow you to see a wide range of species, including the rare Nicaraguan Grackle, Sungrebe and Jabiru, and maybe a rare crake or two if you can get into the hyacinth-covered lagoons. Spectacled Caimans sit along the river banks, and Black River Turtles, Pacific Basilisks and Green Iguanas may all be seen. The surrounding fields and marshes also play host to several difficult-to-find birds, such as the Lesser Yellow-headed Vulture and Pinnated Bittern, and the superb Nicaraguan Seed Finch, while around the two lodges you should look for the Spot-breasted Wren, Grey-headed Dove, and both Common and Great Potoos. Note that the area tends to dry up around March–April.

LA SELVA BIOLOGICAL STATION (OTS)

La Selva is a world-class biological station and nature reserve operated by the Organization for Tropical Studies (OTS). Established in 1953, it encompasses 1,536ha of lowland tropical rainforest, and is probably the most studied rainforest anywhere in the world. Four major tropical life zones are protected here, and within these a remarkable 500,000 species have been recorded, including more than 5,000 plants, 700 trees, nearly 470 birds, more than 200 mammals, 56 snakes, 25 lizards, 500 butterflies and over 300,000 insects, among them 400 ants.

The approach road to the station is a great starting point for the birdwatcher, and this short section of road is a hotspot for countless species early in the morning. You will need to pay an entrance fee and it is advisable to pre-book your arrival, as a naturalist will need to accompany you on the trails. For those not wishing to walk, the area around the restaurant and reception is a real haven for birds, and because of its openness you can scan tree tops and watch flowering shrubs for species such as the Snowy Cotinga, and oropendolas, motmots, woodpeckers and hummingbirds. The Great Curassow regularly walks through this area, and guides are always on hand to point things out.

Over the river bridge and into the rainforest there are endless trails, and the opportunities to see many birds are equally endless. The Bare-necked Umbrellabird is possible, as well as three species of tinamou, and the guides sometimes have a roosting owl or two staked out. There is also plenty of good birding close to La Selva, and several excellent lodges are located not too far away. Wet marshes near the entrance off the main road can hold the Green Ibis and Nicaraguan Seed Finch, while some of the rocky rivers are good for the Fasciated Tiger

Heron. All in all this is a wonderful place in which to spend several days birding, and if you do not want to stay in one of the nearby lodges then La Selva offers basic accommodation.

MONTEVERDE CLOUD FOREST RESERVE

This reserve, situated in the Tilarán Mountains, has for a long time been one of Costa Rica's premier ecotourist destinations. Established in 1972, it now protects more than 14,200ha of middle- and high-elevation cloud forest. The damp and often very wet environment provides constant moisture, which allows many species of plant and tree to flourish. Lichens, orchids and bromeliads are all abundant, and this in turn results in all other wildlife, including numerous bird species, also being very plentiful.

Some of the most spectacular and iconic species can be seen here, including the superb Resplendent Quetzal and the Three-wattled Bellbird; with 14km of forest trails you can have a great time searching for them. A small hummingbird gallery of feeders near the park entrance attracts up to eight species, including the localized Magenta-throated Woodstar. There are many hotels, lodges, shops and places to eat along the approach to this superb area. It should be noted that the road, all 35km of it from the highway, is atrocious; hopefully it will not change, as it is believed that it discourages mass tourism. The nearby Santa Elena Cloud Forest Reserve is certainly worth a visit as it is an excellent example of a habitat with a similar sample of plants and animals as that in the Monteverde, but within a much smaller area of 310ha, including trails ranging from 1km to 5km in length. This is another good site for the Three-wattled Bellbird, and one of the best places to look for the Chiriqui Quail-Dove.

PALO VERDE NATIONAL PARK

This superb park offers something very different once you have had your fill of rainforests and all that is associated with these wet places. Palo Verde is mostly tropical dry forests, and the park concentrates on conserving a vital floodplain, marshes, limestone ridges and seasonal pools. Run by the OTS, it covers some 18,410ha within the Guanacaste region. The OTS runs the Palo Verde Biological Research Station within the park, and visitors can stay here in basic accommodation if they book in advance. There is a ranger station situated right in front of the main lagoon, where thousands of water birds congregate during the dry season. Among the ducks, herons, storks and ibis, you should look out for the Jabiru, Snail Kite and real Muscovy Ducks. The surrounding dry forest can get super hot and quite buggy at times, but a good selection of species hard to find elsewhere make it worth a visit. Great Curassows are not uncommon here, and the Elegant Trogon, Lesser Ground Cuckoo, Banded Wren and White-lored Gnatcatcher are all possible to find. This dry forest, combined with the marshes, grassy fields and scattered trees, can certainly produce an impressive list.

SAVEGRE VALLEY

The Savegre Valley, or as it is better known Los Quetzales National Park or San Gerardo de Dota, gives access to many of the sought-after highland species. Situated in the Central

River at Savegre

Highlands, its habitats encompass misty mountain tops, short paramo and cloud forest literally dripping with mosses and lichens. Hummingbirds thrive here, and many cafes and lodges put out hummingbird feeders to attract these beautiful jewels. From the highway down to several stunning lodges such as Savegre Mountain Lodge and Trogon Lodge, the scenery is simply breathtaking, with huge, moss-covered trees, mountain streams and lovely views of green rolling hills.

Probably the most famous bird in this area is the gorgeous Resplendent Quetzal. This is where it makes its home, and pairs can often nest right around the cabins at a lodge. Plenty of other birds will vie for your attention, especially hummingbirds, nightingale-thrushes, tanagers, warblers, American Dippers, Torrent Tyrannulets and, if you are lucky, the Costa Rican Pygmy Owl and Spotted Wood Quail. It can get quite chilly at the lodges, which are located at around 2,135m, but because of this bird activity does not usually get going until after a nice, filling breakfast. The guides associated with these lodges are superb, and know all the birds and where to find them. They constantly monitor the Resplendent Quetzals, and know which avocado trees they are feeding on and even where they are nesting.

PELAGIC BIRDWATCHING

Costa Rica is not really known for its pelagic birding, but still opportunities exist, with organized trips to Cocos Island offering probably the best chance to see a good variety of species. The open water between the Pacific mainland and the island could in theory produce many different seabirds. Wilson's Petrels should no doubt be present, but depending on the time of year and weather conditions, it may be possible to see the Red-billed Tropicbird, Masked, Nazca and Blue-footed Boobies, Leach's, Band-rumped, Black, Least and Wedge-rumped Storm Petrels, Pink-footed, Wedge-tailed and Galapagos Shearwaters, and Pomarine and Parasitic Jaegers. Apart from the birds there are also great opportunities to see Whale Sharks, Hammerhead Sharks and manta rays, and for divers there is a world of colourful fish and sea life. Anyone interested in visiting the island should check out the following website: www.underseahunter.com

Great Tinamou ■ *Tinamus major* 43cm

DESCRIPTION Largest of the tinamous in Costa Rica, the size of a small turkey. Adults have dark olive-brown upperparts speckled with black. Paler underparts finely barred with black, cinnamon undertail, very short tail, throat whiter, distinctive black eyes and grey legs. **HABITS AND HABITAT** Common resident of low- and middle-elevation rainforest, forest edges and cloud forest at altitudes to 1,500m. Very well camouflaged and easily overlooked if standing still. Most often seen singly or in pairs, and usually on the ground. Call a distinctive set of four downscale, tremulous notes that are far carrying and most easily heard around dusk. **SITES** Widely spread but most easy to see in Carara NP and La Selva OTS.

Black-bellied Whistling Duck ■ *Dendrocygna autumnalis* 53cm

DESCRIPTION Male and female similar with obvious pink bill and legs, grey head, rufous neck, breast and back, black belly and tail, white in the wings most easily seen when flying, and white eye-ring. **HABITS AND HABITAT** Widespread resident of freshwater marshes, ponds and wet pastures. Usually in groups often numberings hundreds. Feeds on aquatic life as well as seeds and even crops, which is why it is often considered a pest by farmers. Call a distinctive high-pitched whistle. **SITES** Easily seen in Palo Verde NP, Caño Negro and many other wetland sites.

Blue-winged Teal ■ *Anas discors* 38cm

DESCRIPTION Smallish dabbling duck. Male has very distinctive white crescent on face and bluish-grey head; rest of body brownish with black undertail and white patch on rear of flanks. Female nondescript brown with light patch on bill base. Both sexes have greyish beaks and yellow legs. In flight shows pale blue patch on upper wing. **HABITS AND HABITAT** Abundant migrant in September–April, in freshwater marshes, ponds, lakes and flooded pastures. Widespread resident of freshwater marshes, ponds and wet pastures. Feeds in shallow water and is easily flushed; flies off in fast and erratic flight. Call a thin, high-pitched 'tsee'. **SITES** Easily seen in Palo Verde NP, Caño Negro and many other wetland sites.

Grey-headed Chachalaca ■ *Ortalis cinereiceps* 51cm

DESCRIPTION Head and upper neck dark grey, rest of body rufous-olive-brown, tail dark brown-black tipped with pale creamy-white, dark grey legs and bill, small red throat-patch, and in flight shows rufous primary flight feathers. **HABITS AND HABITAT** Arboreal and social species often seen in groups of up to 12. Fairly common in lowlands and foothills to 1,100m. Often seen along forest edges climbing through trees and gliding from one tree to another, where it can easily disappear into the foliage. Call a variety of loud, raucous squawks. **SITES** Occurs in many areas, but easy to see at Arenal Observatory Lodge.

Crested Guan ■ *Penelope purpurascens* 89cm

DESCRIPTION Large, mostly dark brown bird with long black tail that is tipped greyish. Neck, breast, belly and back flecked with white streaks and spots, obvious red flap of bare skin on throat, reddish legs and dark grey bill. **HABITS AND HABITAT** Fairly common resident found in forests and forest edges, where pairs and family groups often occur in fruiting trees. Walks easily along branches and despite large size can often be difficult to spot. Found on the Pacific Slope to 1,800m and on the Caribbean Slope to 1,200m. Variety of honking, yelping, whistling calls, plus rustling of wing feathers as it glides in flight. **SITES** Easily seen at La Selva OTS and Arenal Observatory Lodge.

Black Guan ■ *Chamaepetes unicolor* 65cm

DESCRIPTION Mostly glossy black, turkey-like bird, with short tail and obvious red legs. Gorgeous powder-blue facial skin and bright red eyes. Unlikely to be confused with any other similar species. **HABITS AND HABITAT** Arboreal species that is uncommon in highlands and montane forests, from 1,200m to timberline, where it favours evergreens. Can be difficult to see when hidden in tree tops, but does readily comes to feeders at certain lodges. Mostly silent, though occasionally utters deep groan or 'tick' during breeding season. Wings make a loud, whip-cracking rattle as it glides between trees. **SITES** Best site, where 20 or more birds come to feeders, is Bosque de Paz Lodge; also regularly seen in Monteverde Cloud Forest Reserve and Savegre Valley.

Great Curassow ■ *Crax ruber* 91cm

DESCRIPTION Very large and robust with prominent crest and long tail. Male mostly black with clean white lower belly and undertail, long, curly black crest, yellow cere and knob, grey bill and grey legs. Female mostly rufous-brown; tail barred with white, upperparts finely barred with white and black, neck, head and long crest all barred with black and white, yellowish bill and red eyes. **HABITS AND HABITAT** Locally common resident found on forest floor to 1,200m on both the Pacific and Caribbean Slopes. Often seen foraging among dead leaves, where it usually proves shy and walks away once spotted, or climbs fallen trees for safety. Seen in pairs and occasionally small groups of up to six individuals. Call a peculiar high-pitched, descending whistle; male in breeding season gives low, deep resonant boom, repeated several times. **SITES** Found in forested habitat throughout Costa Rica, but regularly seen at La Selva and Arenal.

Least Grebe ■ *Tachybaptus dominicus* 24cm

DESCRIPTION As the name suggests, a small grebe. Overall grey-brown, with paler underparts, dark blackish crown and black throat in breeding plumage. Throat whiter in non-breeding birds. Distinctive bright yellow eyes and slim dark bill separate this from any other grebe in the region. **HABITS AND HABITAT** Common resident found in wide variety of wetlands such as freshwater ponds, lakes, streams, rivers, mangrove swamps and marshes, especially those with plenty of vegetation and cover. Dives underwater to feed on aquatic insects, small fish and frogs. Call a series of rolling rattles or churrs. **SITES** Found in lowlands throughout Costa Rica, although easiest to see in Guanacaste, and Central Valley and northern Pacific. Rare in Tortuguero NP and above 1,500m.

Pied-billed Grebe ■ *Podilymbus podiceps* 33cm

DESCRIPTION Larger than Least Grebe (see p. 17), stockier and mainly brown with darker head and back, and light undertail. Thick, pale bill that in breeding plumage has black band. Also in breeding plumage: black throat and fairly distinct, whitish eye-ring. **HABITS AND HABITAT** Resident found in lowland freshwater ponds, lakes and rivers with vegetation; moves around depending on water levels. Dives underwater to feed on aquatic insects, small fish and frogs, and can often emerge far away from where it initially dived. Call a series of high-pitched rattling notes, 'kee-kee-kee-kee-kee'. **SITES** Widespread but uncommon in suitable habitat to 1,500m.

Wood Stork ■ *Mycteria americana* 102cm

DESCRIPTION Large, predominately dirty white bird with grey head and neck, and heavy, long, decurved bill with droop at tip. Long legs grey-black, and in flight trailing edge to wings, plus tail, show black. Only similar birds in flight are the all-white winged Jabiru (see opposite) and King Vulture (see p. 30), but note length of legs and neck in flight, which separate it from the latter. **HABITS AND HABITAT** Found in both freshwater and saltwater receding shallow pools, where prey such as fish and frogs is easy to catch. Often feeds and roosts communally and can be found in any open wetland habitat. Mostly silent. **SITES** Locally common in dry season in Guanacaste, Gulf of Nicoya and Caño Negro.

Jabiru ■ *Jabiru mycteria* 132cm

DESCRIPTION Huge, white-bodied stork with massive, slightly upturned, sharp-looking black bill. Bare black skin on head and neck, and obvious area of bright red skin on lower neck. In flight wings show completely white. The Jabiru is the tallest flying bird within its range. **HABITS AND HABITAT** Uncommon species found in freshwater marshes and ponds, where it usually occurs alone or in pairs. Feeds on fish, frogs, snakes, turtles and even young caiman. Roosts high in tree tops and builds massive stick nest also high in a tree top. Silent except for bill clacking near nest. **SITES** Only really likely to be seen at Palo Verde NP and in wetlands near Caño Negro.

Green Ibis ■ *Mesembrinibis cayennensis* 55cm

DESCRIPTION Overall plumage dark bronzy-green, which can appear iridescent and almost black or lighter green, depending on the light. Green decurved bill and dull green legs. In flight shows very broad, rounded wings and short tail. **HABITS AND HABITAT** Resident and found locally throughout Caribbean lowlands, preferring wooded swamps, mangroves and forested wetlands. Often in pairs and sometimes small groups. Generally secretive and flies into trees if disturbed, giving a loud rapid, rolling 'kraww-krawwr-kraww-kraww'. **SITES** Easiest to observe from boats and regularly seen on Rio Frio, Rio Sarapiqui and in Caño Negro area.

American White Ibis ■ *Eudocimus albus* 64cm

DESCRIPTION Adults have overall white plumage, but note black wing-tips in flight. Decurved bill and legs are bright red, making it easy to identify. Juveniles, however, are brown above with pale rump and belly, pink-brown bill and grey legs. Adults have pale blue eyes; these are greyer in juveniles. **HABITS AND HABITAT** Locally common and found in both fresh and salt water, favouring tidal mudflats, estuaries and mangrove swamps. Usually seen feeding in flocks and roosts communally in trees, especially mangroves. Call a honking sound, 'urnk-urnk-urnk', most often used in flight. **SITES** Easy to see in Caño Negro and around the Gulf of Nicoya; uncommon elsewhere.

Roseate Spoonbill

■ *Platalea ajaja* 81cm

DESCRIPTION The only large pink bird in Costa Rica. Adult has bare pale greenish-grey head with distinctive spatula-shaped, pale green bill and bright red eyes. Legs are greenish while rest of plumage can vary in intensity from bright red-pink in adults to dull white-pink in juveniles. **HABITS AND HABITAT** Found in shallow fresh and salt water, where its distinctive feeding action involves side-to-side sweeping of the bill, which helps disturb and filter food from the shallow mud. Mostly silent, but in flight gives a nasal 'kuh-kuh-kuh'. **SITES** Fairly common in Gulf of Nicoya and Caño Negro. Rarer on Caribbean side.

Fasciated Tiger Heron ■ *Tigrisoma fasciatum* 64cm

DESCRIPTION Medium-sized, grey-olive heron with stocky neck and short legs. At close range all upperparts look finely barred. Crown dark grey and black line runs down either side of neck, bordering white stripe with rufous centre running from throat to upper breast. Bill mostly grey with yellow base to lower mandible. Yellow eyes and dull olive-grey legs. Immature birds rufous-brown all over, with dark barring. **HABITS AND HABITAT** Uncommon and rarely seen away from fast-running, rocky streams and rivers, where it often sits motionless on a rock or boulder. Feeds mostly on fish. Call a deep 'kwok'. **SITES** Regular along Rio Sarapiqui, and streams near Arenal and Caribbean Slope to 800m.

Bare-throated Tiger Heron ■ *Tigrisoma mexicanum* 78cm

DESCRIPTION Medium–large heron with stocky neck and short legs. Similar to smaller Fasciated Tiger Heron (see above), but easily identified by yellow throat, which is present on both adults and juveniles. Chestnut-and-white stripe runs from lower throat to upper belly. Rufous lower belly, dull olive-grey legs, orange-yellow eyes, and grey bill with more extensive yellow on lower mandible than in Fasciated. **HABITS AND HABITAT** More widespread than Fasciated Tiger Heron and less specific in habitat preference, being found in pretty much any type of wetland, including marshes, ponds, rivers, mangroves and even watery ditches. Feeds mostly on fish, frogs and crabs. Call a deep 'kworr-kworr-kworr'. **SITES** Easily seen at Palo Verde, and can be photographed during boat trips on Rio Tárcoles.

Agami Heron ■ *Agamia agami* 75cm

DESCRIPTION Very attractive, medium-sized heron with incredibly long, thin bill and neck. Overall colours range from glossy green of wings, back, head and tail, to rich chestnut belly and upper sides of neck. White-and-chestnut stripe runs from throat down to upper belly, and sides of neck are adorned with beautiful blue-white, filament-like feathering. Reddish-brown eyes, and bill mostly grey with olive-green lower mandible. Dark legs. **HABITS AND HABITAT** Uncommon to rare resident in lowland forests. Very solitary and secretive heron found in small forest streams, rivers with thick vegetation and sometimes mangrove edges. Can be very difficult to spot as it prefers shady cover from where it hunts small fish and frogs. Mostly silent. **SITES** One of the most frequent sites in which it is recorded is La Selva OTS.

Boat-billed Heron ■ *Cochlearius cochlearius* 51cm

DESCRIPTION Medium-sized, dumpy heron with large head, short neck and unusual broad, shoe-like bill. Upperparts mostly pale grey; black crown, long crest and back of neck. White face with white patch on forehead. Underparts pale buffy-pink, becoming darker on lower belly. Huge dark eyes and grey-green legs. **HABITS AND HABITAT** Rarely seen during the day unless found in a communal roost in trees or bushes often overhanging water. This strange-looking nocturnal bird uses its large eyes to help it find frogs and other amphibians, fish and crabs. Mostly found in or around mangroves, estuaries and forested rivers. If disturbed gives a loud snapping of the bill, and calls a deep 'kwaaa-kwaa-kwaa-kwa-kwa-kwa'. **SITES** Best seen from boats on Rio Tárcoles and Caño Negro.

Black-crowned Night Heron ■ *Nycticorax nycticorax* 64cm

DESCRIPTION Medium-sized, mostly grey, stocky, short-necked heron with paler grey underparts, black back and crown with long white plume. Large red eyes, shortish yellow legs and black bill. Immatures have yellower bill than adults, and are brown-grey with large white spots and speckles. **HABITS AND HABITAT** Nocturnal species that during the day roosts communally near or overhanging water. Found in marshes, swamps, mangroves, estuaries and edges of rivers, ponds and lakes. Feeds on fish and frogs. At night when flying to feed, utters harsh single 'kwork'. **SITES** Can be seen on boat trips in daytime on Rio Tárcoles and Caño Negro.

Yellow-crowned Night Heron ■ *Nyctanassa violacea* 61cm

DESCRIPTION Similar in size and shape to Black-crowned Night Heron (see above). Mostly pale grey with obvious broad black-and-white head pattern and prominent white crest. Neck, breast and belly pale grey, and wings and back have dark grey-black feathers with silver-grey edging, giving a mottled appearance. Orange eyes and olive legs. Brown-grey immatures can be distinguished from previous species by smaller white spotting and all-dark bill. **HABITS AND HABITAT** Not quite as nocturnal as previous two species, and can often been seen on estuaries, mudflats and mangrove edges during the day. Prefers salt-water habitats, but can also be found along river and pond edges. Feeds mainly on crabs. Gives loud, deep 'kwok' notes at night. **SITES** Numerous and easy to see on Rio Tárcoles.

Green Heron ■ *Butorides virescens* 44cm

DESCRIPTION Small, generally dark-looking heron. When seen well shows glossy green wings and back, contrasting with rich chestnut neck fading to grey belly, dark glossy green-black cap and bright yellow legs. Eyes [illegible] bright yellow, and bill has black upper and yellow lower mandibles. **HABITS AND HABITAT** Relatively common heron found near any body of water to 1,800m, including salt-water estuaries, ponds, marshes, rivers and wet pastures. Often raises rather shaggy crest when alert, and feeds low along shallow water edges. Call a loud 'kwark' in flight. **SITES** Easily seen at many sites, including Rio Tárcoles, Caño Negro, Rio Sarapiqui and Palo Verde.

Western Cattle Egret ■ *Bubulcus ibis* 51cm

DESCRIPTION Smallest all white-looking heron. Told from all other white herons except much larger Great Egret (see opposite) by yellow-orange bill and shorter neck. In breeding plumage shows buffy orange on crown, chest and back. Immatures and non-breeding plumage dull white, with duller bill. Yellow eyes, and black legs and feet. **HABITS AND HABITAT** Abundant and (as its name suggests) most often seen around livestock, where it catches insects that are flushed up from the feeding animals. Often seen in grassland and arable fields in large feeding groups. Considered abundant and also forms huge communal roosts. **SITES** Many sites, but easy to see around Gulf of Nicoya and Rio Tárcoles areas.

Great Blue Heron ■ *Ardea herodias* 117–132cm

DESCRIPTION Largest grey-looking heron in Costa Rica. Head appears white at a distance, but note black stripe extending from side of crown to short black crest. Long neck grey with central white stripe edged with black going down to long, shaggy grey plumes on upper breast. Rufous thighs diagnostic, while legs are dark grey, and bill and eyes are pale yellow-orange. Flies with slow wingbeats. **HABITS AND HABITAT** Common migrant; in September–May can be found in open wetlands and pastures, as well as rivers, ponds and estuaries. Feeds on fish, amphibians, insects and even small mammals. When disturbed gives loud, harsh 'kwark'. **SITES** Easy to photograph from boats on Rio Tárcoles, and at Palo Verde.

Great Egret ■ *Ardea alba* 100cm

DESCRIPTION Largest of all-white herons, and looks similar in size to Great Blue Heron (see above). Obvious yellow bill and long black legs help distinguish it from all herons except the much smaller Western Cattle Egret (see opposite). Leisurely, slow wingbeats in flight. Pale yellow eyes. **HABITS AND HABITAT** This often solitary heron is abundant and can be found near any body of salt or fresh water. Sometimes occurring in small groups, it can be seen in association with other herons at communal roosts. Residents are joined by migrants in September–May. Feeds on fish, frogs and other amphibians, and is usually silent. **SITES** Easy to see at many wetlands, but there are good photographic opportunities from boats on Rio Tárcoles.

Tricolored Heron

■ *Egretta tricolor* 66cm

DESCRIPTION Medium-sized, slender heron with long neck, thin bill and white crest plumes. Back and wings slate-grey, head and neck mostly slate grey-blue with neck stripe, and belly a contrasting white that is obvious in flight. Lower neck shows chestnut. Yellowish eyes, creamy-grey bill and yellow legs. **HABITS AND HABITAT** Mostly solitary, but in suitable habitat such as estuaries, freshwater ponds, marshes and rivers, several can be seen together. Feeds in same way as Snowy Egret (see opposite), by stirring water with feet to flush small fish. Common migrant in September–May. Usually silent. **SITES** Great photographic opportunities from boats on Rio Tárcoles.

Little Blue Heron

■ *Egretta caerulea* 60cm

DESCRIPTION Medium-sized heron that looks overall slaty-blue with purplish tones, and has long blue plumes on back. Light blue bill with distinct black tip a third of the entire bill length. Dull blue-green legs. Immature birds can be confused as they appear white or dirty white, but bill is dull two tone like in adult, unlike bills of other similar white egrets. **HABITS AND HABITAT** Common and widespread migrant in September–April to 1,500m. Found in almost any wetland habitat, including ponds, streams, rivers, lakes, mangroves, estuaries and wet pastures. Often feeds alone, although small groups are sometimes seen. Mostly silent. **SITES** Gulf of Nicoya; best photographic opportunities from boats on Rio Tárcoles.

Snowy Egret ■ *Egretta thula* 61cm

DESCRIPTION Medium-sized pure white heron, most easily identified by black legs with bright yellow feet if they are visible. No other white egret in Costa Rica has yellow feet. Straight bill is black, and area between bill and eye (lores) is yellow. Eyes are also yellow, and breeding plumage birds often have long white plumes extending from nape, breast and back. **HABITS AND HABITAT** Common resident found in almost any type of wetland to 1,500m. Very lively and often seen dashing back and forth in search of small fish. Can also be seen using one foot at a time to stir up mud in search of food. Mostly silent. **SITES** Gulf of Nicoya, Rio Tárcoles and Caño Negro, among others.

Brown Pelican ■ *Pelecanus occidentalis* 105–110cm

DESCRIPTION Very large bird with light grey back and dark belly. Head and neck variable depending on breeding condition, but head mostly white, back of neck dark brown and throat white. Huge diagnostic bill has dark lower mandible, and creamy or blue upper mandible. Eyes are pale. **HABITS AND HABITAT** Common on Pacific coast; rarer on Caribbean coast. Found mostly close to shore and in estuaries, where groups can be seen either feeding or roosting together. Often seen plunge diving just offshore, where it mostly feeds on anchovies and sardines. Also flies in lines like geese, when it uses its 2.4m wings to mostly glide. Silent. **SITES** Shores of Gulf of Nicoya and Rio Tárcoles river mouth are good places to see it.

Magnificent Frigatebird ■ *Fregata magnificens* 90–114cm

DESCRIPTION Largest of the frigatebirds, with 2.1–2.4m wingspan. Overall appearance is all black, with immatures and females showing varying amounts of white on belly, throat and inner wing where it joins the body (axillaries). Males have bright red throat-patch, which can be inflated like a balloon. Long, hook-tipped bill varies from dark grey to horn. **HABITS AND HABITAT** Fairly common on both Caribbean and Pacific coasts, where its distinctive shape can be seen as it patrols offshore following fishing boats or chasing other seabirds. Also hunts schools of fish and can be seen perched near shore on tree tops. Silent away from nesting colonies. **SITES** Anywhere along coasts where fishing boats are present. Regular at Rio Tárcoles and Punta Leona.

Neotropic Cormorant

■ *Phalacrocorax brasilianus* 60–72cm

DESCRIPTION Adults mostly glossy black or brown with longish tail, shortish neck, white edges to yellowish throat-patch, and thin, hooked darkish bill. Eyes are blue-green. Immatures browner than adults all over, and slightly paler on underparts. In breeding plumage adult acquires white plumes on sides of head. **HABITS AND HABITAT** The most common cormorant of Costa Rica, found mostly in shallow freshwater ponds, rivers and marshes to 1,500m. Also in estuaries and on shoreline, where it can be seen fishing and flying in groups, or perched on tree tops or logs. Generally silent but grunts at roosts. **SITES** All around Gulf of Nicoya, Rio Tárcoles and Caño Negro. Breeds on Arenal Lake.

Anhinga ■ *Anhinga anhinga* 86cm

DESCRIPTION Similar to cormorants but easily told by its very long, skinny neck, small head, long, sharp, pointed bill and long, buff-tipped tail. Adult male glossy black with silvery feathers on wings, back and sides of head. Red eyes. Female similar to male, but has fawn-brown head and neck to upper breast. **HABITS AND HABITAT** Widespread resident of slow-flowing rivers, lakes and mangroves. Perches with head and bill pointing upwards and wings and tail spread. Often swims with just its neck and head out of water, which is why it is also known as 'snakebird'. Usually silent. **SITES** Easy to see close up on boat trips on Rio Tárcoles and Caño Negro.

Turkey Vulture ■ *Cathartes aura* 76cm

DESCRIPTION Mostly black except for bare red head. Hind-neck bluish on resident birds, red on migrants. Pale horn-cream bill and short, yellow-white legs. Can be identified in flight by slightly up-tilted wings, with pale silvery-grey flight feathers on underwing and longer tail than that of similar Black Vulture (see p. 30). **HABITS AND HABITAT** Common resident to 2,000m, with many additional migrants in September–October and January–April, mostly on Caribbean side. Feeds on carrion in association with other vultures. Silent. **SITES** Can be seen pretty much anywhere and easily visible along roadsides.

Black Vulture ■ *Coragyps atratus* 64cm

DESCRIPTION Entirely black with bare, wrinkled black skin around head and pale whitish legs. Easily identified in flight by very short, square tail and fairly obvious white patch at end of wings. Soars on flatter wings than the other two 'black' vultures, and also flaps rapidly when not gliding. **HABITS AND HABITAT** Very common resident throughout Costa Rica, including in cities and towns. Less frequent over forests or above 2,000m. Feeds on carrion, refuse and fruits, and also preys on young hatchling turtles. Silent. **SITES** All over Costa Rica, but easily seen at carcasses along roadsides and at refuse dumps.

King Vulture

■ *Sarcoramphus papa* 81cm

DESCRIPTION Biggest and 'king' of Costa Rican vultures. Adults a distinctive black and white. Flight feathers and tail black, strongly separated from light grey-white body and wings. Very obvious white eyes set on bare wattled head, which is a mixture of red, blue, orange and yellow. Orange bill and grey legs. Immatures are brown. **HABITS AND HABITAT** Resident and uncommon in lowlands below 1,200m. Fairly solitary, with 3–4 occasionally seen soaring together over mature forests. Can detect carrion from vast distances. **SITES** Regularly seen above Villa Lapas, Carara and La Selva OTS.

Western Osprey ■ *Pandion haliaetus* 58cm

DESCRIPTION Fairly large, long-winged bird that is brown above and mostly white below. Head always looks white from a distance, but close up it shows dark line through eye to back of neck. White underwing shows chequered flight feathers and black wing-tips. Bright yellow eyes, long, hooked black bill and grey legs. **HABITS AND HABITAT** Common visitor in September–April, when it can be seen along the coasts as well as around lakes and even trout ponds. Perches on exposed branches and often seen hovering over water before plunge diving to catch a fish, its predominant prey. Call a high-pitched 'tweek-tweek'. **SITES** Easy to see along coast of Gulf of Nicoya, and from boats on Rio Tárcoles and Caño Negro.

White-tailed Kite ■ *Elanus leucurus* 41cm

DESCRIPTION Small, long-winged, bright white hawk. Body, head and tail pure white, and upperparts pale grey with darker wing-tips and black shoulders. Underwing shows greyish wing-tips and dark spot under bend in wing. With good views, dark line through eye and large, bright red eyes can be seen. Yellowish legs. **HABITS AND HABITAT** Common resident in lowland open areas to 1,500m. Often seen hovering over grassland and farmland, as well as city edges and even airports, from where it drops onto prey such as small mammals, lizards and insects. Call a sharp, high-pitched 'kwik-kwik-kwik'. **SITES** Can be seen anywhere in suitable habitat, but there are always a few around San José airport.

Pearl Kite ■ *Gampsonyx swainsonii* 23cm

DESCRIPTION Smallest raptor in the Americas. Mostly white underparts; back of head, neck, wings, uppertail and back slaty-grey; forehead and sides of face white with peachy-buff wash, and rufous flanks. Thin black collar on sides of neck. Reddish eyes, black bill and yellow legs. **HABITS AND HABITAT** Uncommon but increasing on the Pacific side since it was first recorded in Costa Rica in the 1990s. Often seen perched on wires near open pastures and fields with scattered trees. Flies swift and fast like a falcon as it hunts small birds, lizards and grasshoppers. Mostly silent. **SITES** Fairly reliable in suitable open habitat around San Isidro, La Esquina and Osa Peninsula.

Swallow-tailed Kite ■ *Elanoides forficatus* 58cm

DESCRIPTION When perched shows white head and underparts, and clean black upperparts, wings and tail. A truly elegant bird, in flight displaying diagnostic long, forked tail unique among raptors, combined with white underparts and underwing with black flight feathers. Red eyes, black bill and grey legs. **HABITS AND HABITAT** Common breeding bird in November–March. Mostly seen in flight, soaring effortlessly low over hills, forests and mountains to 2,000m in search of insects, snakes and lizards that it then eats in flight. Mostly silent but can utter an ascending series of high-pitched notes. **SITES** Among others, Savegre Valley, San Isidro, Carara and Osa Peninsula.

Double-toothed Kite ■ *Harpagus bidentatus* 35cm

DESCRIPTION Medium to small bird with slate-grey head and upperparts. Upper breast bright rufous, and lower breast and belly white with distinct rufous barring. Undertail white (obvious in flight), with four broad dark bands. Throat pure white, broken by thin black line from bill. Orange eyes, yellow bill with black tip and yellow legs. **HABITS AND HABITAT** Fairly common resident to 1,500m. Often seen soaring over forest edges, where it looks like an accipiter. Often occurs near monkeys and army ants, searching for lizards and insects that get disturbed. Call varies from 'tsip-tseep-tseep-tseep' to high-pitched whistle, 'see-weeeet'. **SITES** May be seen in any patch of suitable forest, such as Carara, Braulio Carrillo or La Selva.

Plumbeous Kite

■ *Ictinia plumbea* 36cm

DESCRIPTION Overall looks completely slaty-grey, paler on underparts. Flying birds show distinctive rufous patch on outer wing (primaries), and black tail with two thin white bands (not always easy to see). Red eyes, black bill and orange legs. Juveniles brown and streaky, but note long wings. **HABITS AND HABITAT** Fairly common breeding migrant in February–September, with added migrants early and late in that period. Found along Caribbean and Pacific coast lowlands, preferring open areas, forest edges, rivers and mangroves. Often soars in groups on flat, straight wings, and likes to perch on dead tree tops. Catches lizards, snakes and many insects. Call when breeding a shrill 'szweeeeoo'. **SITES** Fairly easy to see at Caño Negro and Tárcoles mangroves.

Snail Kite

■ *Rostrhamus sociabilis* 43cm

DESCRIPTION Overall dark slaty-grey-black, contrasting strongly with white base to uppertail and white undertail-coverts, easily seen when in flight. Wings broad and rounded. If seen close up shows very long, hooked bill that is mostly red with black tip; red eyes and long orange legs. **HABITS AND HABITAT** Locally common in preferred habitat of lowland freshwater marshes, where it hunts low in search of snails, which form most of its diet and are the reason for the long, sharp bill. Often sits low on fence posts or bushes, and may congregate in large numbers. Population fluctuates according to water levels. Mostly silent except when breeding. **SITES** Best sites are Caño Negro and Palo Verde NP.

Common Black Hawk ■ *Buteogallus anthracinus* 56cm

DESCRIPTION Overall slaty-black, medium-sized raptor with bright yellow cere and lores, and black-tipped bill. One broad white band across centre of tail. In flight tail-band is obvious, as are broad wings. Dark eyes and yellow legs. Juveniles brown with heavy streaking on chest and barred undertail. **HABITS AND HABITAT** Common resident along both Caribbean and Pacific coasts. Almost always near water, including mangroves, mudflats and marshes, where it feeds on crabs, lizards and frogs. Birds around mangroves were considered a separate subspecies, the Mangrove Black Hawk, but this is now considered incorrect and only the Common Black Hawk occurs in Costa Rica. Call a series of high-pitched whistles. **SITES** Excellent views can be had from boat trips on Rio Tárcoles.

Roadside Hawk

■ *Rupornis magnirostris* 38cm

DESCRIPTION Upperparts slate-grey with completely pale grey head and upper breast. Underparts pale with rufous barring, and tail has 4–5 black bands. In flight shows distinctive rufous patch in primaries. Pale yellow eyes, yellow cere, yellow bill with black tip, and yellow legs. **HABITS AND HABITAT** Fairly common and widespread resident in secondary growth, open areas and scattered woodland to 1,500m. Perches low in trees, and on fence posts and telephone poles, from where it hunts small mammals, lizards and insects. Vocal raptor uttering high-pitched 'kweee-kweee'. **SITES** May be seen in any suitable lowland area. Particularly easy to see along roadsides in Guanacaste.

White Hawk

■ *Pseudastur albicollis* 58cm

DESCRIPTION Striking, predominately snow-white raptor with black primaries and single black band in white tail. In flight shows broad, mostly white underwing and short tail. Dark eyes, grey cere and bill, and yellowish legs. **HABITS AND HABITAT** Fairly common resident in and around lowland forests on both Caribbean and Pacific Slopes to 1,200m. Rarer in Gulf of Nicoya. Favours forests and forest edges, often perching in trees on exposed branches, from where it glides down onto prey. Readily seen soaring over forests, where it gives a harsh 'shweeea' call. **SITES** A good place to see it is around the dam of Arenal Lake.

Grey Hawk ■ *Buteo plagiatus* 42cm

DESCRIPTION Overall light grey upperparts including head and wings; light underparts finely barred grey. In flight shows white underside with bands on tail; this is more obvious on black uppertail with one obvious white band. Brown eyes, yellow cere, bill with black tip and yellowish legs. **HABITS AND HABITAT** Locally common resident in lowlands of north-west down to Rio Tárcoles. Favours forest edges and savannah, where it often perches in the open, or on dead branches or telephone poles. Feeds on lizards and insects, flies fast and regularly soars. Call a high-pitched 'pyeeeer-yeer-yeer'. **SITES** Can be seen along roadsides in Guanacaste and Caño Negro areas, and in Rio Tárcoles area.

Broad-winged Hawk

■ *Buteo platypterus* 42cm

DESCRIPTION Immatures' upperparts dark brown including head, which shows distinct white line above eye; throat pale with distinct dark malar stripe. Breast heavily streaked brown, and short tail evenly barred black and white. In flight pale underneath with black tips to flight feathers. Brown eyes, yellow cere and legs, and grey bill tipped black. **HABITS AND HABITAT** Common winter resident and migrant in September–May, when majority of birds are immatures. Favours open areas and forest edges to 2,000m. On migration can be seen in huge flocks numbering hundreds or thousands. Perches at mid-height in trees, from where it swoops onto small mammals, lizards and insects, and often seen soaring. Call a drawn-out whistle, 'peeeeeew'. **SITES** May be seen anywhere.

Short-tailed Hawk ■ *Buteo brachyyurus* 41cm

DESCRIPTION Upperparts and head brown-slaty-grey; underparts pure white from throat down to undertail. Tail finely barred with black terminal band. White forehead and lores, brown eyes, yellow cere, greyish bill and yellow legs. The tail is not, in fact, particularly short. Also seen in rarer dark form, in which white areas of underwing and body are replaced by dark brown. **HABITS AND HABITAT** Fairly common and local resident to 2,000m, and migrant in September–May. Most frequently seen soaring over open areas and forest edges, and not often seen perched. Feeds on small birds, lizards and snakes. Call a piercing, shrill 'shreeeeeea'. **SITES** Sarapiqui around Selva Verde is a good site.

Dark form

Zone-tailed Hawk ■ *Buteo albonotatus* 52cm

DESCRIPTION Turkey Vulture (see p. 29) mimic and surprisingly similar in flight, but has pale band on undertail, pale chequered pattern on underwing, feathered (not bald) head, obvious bright yellow legs and cere, and bill with black tip. Long winged and long tailed. Immatures difficult to identify, but mostly brown. **HABITS AND HABITAT** Uncommon and local resident, and winter migrant, found in lowlands and foothills to 1,500m. Soars and rocks on slightly upturned wings (same as Turkey Vulture), from where it then surprises unsuspecting prey by dropping straight down on it. Call a high-pitched 'kree-kree'. **SITES** Often and regularly seen near Selva Verde Lodge and generally in Sarapiqui area.

Sunbittern ▪ *Eurypyga helias* 46cm

DESCRIPTION Resembles a heron, but has horizontal not upright stance. Cryptic plumage a combination of grey, orange and black barred upperparts, orange-buff neck and breast, white throat and belly, and grey tail with two black bars. Head black with white line above and below the eye. Long, straight, orange-red bill and short, bright orange legs. Bright red eyes. Best known for fantastic sunburst pattern on upperside of broad, rounded wings when open. **HABITS AND HABITAT** Uncommon and local resident on lowland fast-flowing streams on Caribbean and South Pacific Slopes to 1,500m. Sometimes around ponds and wet forests, where it walks slowly in search of insects, frogs and small fish. Song a long, two-second whistle, 'pweeeeeeeeeeee'. **SITES** Streams at Arenal Observatory Lodge, Selva Verde and Monteverde.

Sungrebe ▪ *Heliornis fulica* 28cm

DESCRIPTION Looks like small, slender-bodied grebe, but has distinctive black-and-white striped head and neck, with long, wax-red bill on breeding female, orange cheeks and buff-brown upperparts. Tail black with brown band on tip, and brown eyes. Male and non-breeding birds have dull horn-coloured bill. Legs if seen out of water an amazing striped black-and-green colour. **HABITS AND HABITAT** Uncommon; sits low in water and difficult to spot, especially as it likes to conceal itself in thick vegetation and mangroves. Found on slow-moving rivers and streams with overhanging cover. Swims with clockwork motion and plucks insects from surface, as well as small frogs. Mostly silent. **SITES** Best seen on boat trips on Rio Sarapiqui and Caño Negro; occasionally from La Selva bridge.

Grey-necked Wood Rail ■ *Aramides cajanea* 38cm

DESCRIPTION Looks like large chicken with long, bright red legs, longish yellow bill and bright red eyes. Olive-brown back, wings and tail, grey head and long neck, rich chestnut breast, black lower belly and undertail, and white throat. **HABITS AND HABITAT** Locally common and widespread resident of wet tropical lowlands to 1,500m. Usually found on forest floor near water, where it walks around cocking its tail in search of insects, small reptiles and even fruiting berries. Shy, preferring to be undercover, and runs off rapidly when approached. Often in pairs. Loud and noisy call, 'koor-kak – koor-kak – kor-kak-kor-kak-ka-ka-ka'. **SITES** Easy to see at La Quinta Country Inn, Caño Negro and Rio Sarapiqui.

Common Gallinule ■ *Gallinula galeata* 35cm

DESCRIPTION Looks like small dark duck, but swims with nodding head. Mostly slaty-black with obvious white stripe along sides near closed wing, white undertail, and red frontal shield and bill, the latter with yellow tip. Out of water the long legs and long toes are seen to be yellow-green. **HABITS AND HABITAT** Fairly common, widespread resident in lowlands to 1,500m. Found in ponds, rivers and marshes with slow-flowing water and lots of vegetation. Readily climbs up reeds and eats leaves, seeds, insects and small creatures. Most frequently heard call a loud 'bipp'. **SITES** May be found on any pond, but easy to spot at Palo Verde.

American Coot

■ *Fulica americana* 37cm

DESCRIPTION Very duck-like but completely black with obvious white shield and bill, with dark spot near tip and maroon at top of shield. Also pale whitish line on edge of tail. Bright red eyes and greenish-yellow legs. **HABITS AND HABITAT** Fairly common migrant and winter visitor in October–April in lowland marshes, lakes and ponds. A few birds resident all year. Often in groups on open water, where birds dabble and dive for algae, grass shoots and other plants. Runs across water. Mostly silent. **SITES** Lake Arenal, Palo Verde and Caño Negro, but worth checking any ponds for it.

Limpkin

■ *Aramus guarauna* 66cm

DESCRIPTION May be mistaken for small heron or ibis, but close inspection shows an overall brown-grey bird covered with varying amounts of white flecks. Long, slightly decurved, yellow-grey bill and long, dull greenish legs. Slow, flapping flight with deep down-strokes. **HABITS AND HABITAT** Local and uncommon over much of Costa Rica to 1,500m. Fairly common in marshes at Palo Verde and Caño Negro, where it walks around singly looking for apple snails, its main diet. Roosts and sits in open trees, where it often gives a loud 'kraowww-kraow'. **SITES** Two reliable sites are Palo Verde and Caño Negro.

Double-striped Thick-knee ■ *Burhinus bistriatus* 50cm

DESCRIPTION Mostly streaky-brown with obvious thick white stripe over eye bordered by black. Huge yellow eyes and long yellow legs set it apart from other shorebirds. Large head and upright stance with short tail also distinctive. Short, thick bill half yellow with black tip. In flight shows striking black-and-white patch in outer wing. **HABITS AND HABITAT** Locally common resident in lowlands, but well camouflaged and most active at night, so easily overlooked as it sits in the shade motionless during the day. Found in pastures and grassland, as well as mangroves in the north-west. Calls at night, 'priip-prip – prip-pip-pip-pip-pip-pip'. **SITES** Rio Tárcoles, Hacienda Solimar and Guanacaste area.

Black-necked Stilt ■ *Himantopus mexicanus* 38cm

DESCRIPTION Long, slender shorebird with predominately black upperparts and white underparts. Long, thin neck, small, rounded head with long, needle-like, straight black bill, and very long, pink-red legs. Red eyes. In flight shows striking black wings and white body, with long legs protruding well beyond tail. **HABITS AND HABITAT** Common resident and migrant to 1,500m. Found in shallow saltwater and freshwater marshes, tidal flats, estuaries and ponds. Often occurs in groups, actively plucking insects from the surface or probing mud for molluscs. Excited call, 'trip-trip-trip-trip-trip-trip-trip'. **SITES** Many coastal areas including Gulf of Nicoya, Rio Tárcoles and Caño Negro.

Southern Lapwing ■ *Vanellus chilensis* 36cm

DESCRIPTION Fairly large-looking plover with head and upperparts grey-brown with bronze patch on shoulders. Upper belly and chest black, lower belly and undertail white, rump white and tail black. Black forehead and throat edged with white, and long, thin black crest feathers extend over crown and behind head. Pink bill with black tip, red eye-ring, red eyes and long pinkish legs. **HABITS AND HABITAT** Has become a fairly common resident since it was first recorded in 1997. Now established on both Caribbean and Pacific lowlands to 800m. Favours open, grassy areas near water. Alarm call a harsh 'keee-keee-keee-keee'. **SITES** Seen on boat trips upriver on Rio Tárcoles.

Semipalmated Plover ■ *Charadrius semipalmatus* 18cm

DESCRIPTION Small, chunky plover and the only plover in Costa Rica with orange legs. Also half orange and black-tipped bill. Sandy-brown upperparts, tail, nape and crown, white forecrown, white throat and collar, and small white line above eye. Underparts white, and narrow black or brown breast-band. **HABITS AND HABITAT** Common migrant through September–May, mostly along Pacific coast, less so on Caribbean coast. Found on tidal flats, shingle and sand beaches, mangroves and salinas, and sometimes ponds. Run-and-stop feeding action, during which it picks small molluscs, crustaceans and worms from the surface. Flies in small flocks. Call a soft 'tweep' often given in flight. **SITES** Anywhere along the coasts, including Gulf of Nicoya and Rio Tárcoles.

Killdeer ■ *Charadrius vociferus* 25cm

DESCRIPTION Biggest of 'ringed' plovers, but easily identified by two black breast-bands. Upperparts sandy-brown; underparts white, including throat and collar, forecrown and line behind eye. In flight note chestnut rump-tail with black-and-white terminal band. Pink legs, bright red eye-ring and black bill. **HABITS AND HABITAT** Fairly common lowland migrant found on farmland, estuaries, ponds and even grassy lawns. Can gather in small groups and feeds on insects, crustaceans and worms. Very noisy and flies around giving distinctive alarm call. Call is 'kill-deer – kill-deer – kill-deer – kill-deer'. **SITES** Good places to see it include boat trips on Rio Tárcoles and Hacienda Solimar.

Northern Jacana ■ *Jacana spinosa* 23cm

DESCRIPTION Black head, neck and breast, chestnut back and maroon belly. Closed wings chestnut; flight feathers yellow-green. Yellow shield on forehead, yellow bill, pale blue cere and long, greenish legs with exceptionally long toes. Completely different brown-and-white juveniles often mistaken for shorebirds, but have long toes. **HABITS AND HABITAT** Common, widespread resident of lowlands to 1,500m. Found in wet pastures, marshes, ditches, ponds and estuaries, all with aquatic vegetation, especially water hyacinth and water lilies. Walks and runs jerkily over surfaces of water plants, where it catches insects and small fish. Call a sharp 'zip-zip-zip-zip-zip-zip-zip-zip-zip-zip-zip'. **SITES** Easy to see at Rio Tárcoles and Caño Negro, and in La Selva and La Esquinas areas.

Solitary Sandpiper ■ *Tringa solitaria* 21cm

DESCRIPTION Small, often inconspicuous, dark-backed sandpiper. White underparts, brownish head and neck, and straight, medium-length dark bill. Most distinct feature is bold white eye-ring. Wings and back can look speckled, but in flight wings and rump show dark. Legs dull grey-green. **HABITS AND HABITAT** Uncommon but widespread migrant in August–May in freshwater habitats such as ponds, marshes, river edges and flooded fields. Often alone or in small numbers feeding in shallows, where it bobs its head and tail. Flies off low and fast if disturbed, and calls a soft 'tewit-twit'. **SITES** Ponds in Guanacaste, but may be found in any suitable habitat.

Spotted Sandpiper ■ *Actitis macularius* 19cm

DESCRIPTION Fairly small, with olive-brown upperparts and white underparts with brownish smudge on sides of breast. In breeding plumage has spotted underparts, but it is unlikely to be seen like this during its period in Costa Rica. Constantly bobs body. White line above eye, yellowish bill and dull yellow legs. **HABITS AND HABITAT** Common and widespread migrant in August–May, and found in just about any wet habitat to 2,000m. Mudflats, ponds, river edges and mountain streams, where it catches insects, crustaceans and small fish. Can be seen in big flocks going to roost. Call is 'pweet-weet'. **SITES** Many can be seen on boat trips on Rio Tárcoles.

Least Sandpiper ■ *Calidris minutilla* 15cm

DESCRIPTION Smallest American shorebird, which is mostly brownish above with finely streaked brown breast and white belly. Thin white line above eye and short, straight black bill; key feature separating it from other small sandpipers is dull yellow legs. **HABITS AND HABITAT** Common migrant in August–May in both freshwater and saltwater sites. Habitats include estuaries, rivers, ponds, marshes and shorelines. Often in flocks and readily seen alongside other shorebirds, where it can be observed searching for small insects and crustaceans. Call a high 'treeep'. **SITES** Easy to see on Rio Tárcoles and Caño Negro.

Black Skimmer ■ *Rynchops niger* 48cm

DESCRIPTION Huge and tern-like with massive red-and-black bill, which is unique as upper mandible is much shorter than lower mandible. Black upperparts and long wings with white trailing edges, and long black tail with white edge. Bright white underparts, and white face and forehead. Black eyes and red legs. **HABITS AND HABITAT** Fairly localized migrant in September–May along Gulf of Nicoya; rarer summer resident on Pacific coast. Found in coastal lagoons, sand banks and estuaries, often among terns and gulls. Flies with slow, shallow wingbeats and skims lower mandible through water in search of fish. Mostly silent. **SITES** Lagoons and Salinas along Gulf of Nicoya. Regular at estuary mouth on Rio Tárcoles.

Laughing Gull ■ *Leucophaeus atricilla* 40cm

DESCRIPTION Medium-sized gull with medium to dark grey back and wings. Wing-tips black and whitish head smudged with grey. White underparts and tail. Distinct white eye-crescents especially in birds in breeding plumage with sooty hood. Heavy, blackish-deep red bill and legs. Immatures have broad black tail-band. **HABITS AND HABITAT** By far the most common gull in Costa Rica, frequenting both Caribbean and Pacific coastal areas. Favours mudflats, estuaries and shorelines, where it often associates with terns. Flies lightly and gracefully, and being a scavenger mostly feeds on offal and scraps. Call a 'ha-ha-ha-ha'. **SITES** Beaches and coastal habitats. Easily seen at mouth of Rio Tárcoles.

Band-tailed Pigeon ■ *Patagioenas fasciata* 35cm

DESCRIPTION Large, mostly grey pigeon with distinctive white collar and bright yellow bill. Hind-neck below collar is greenish, while head, neck and breast have a purplish wash, and lower belly and undertail are white. Tail grey with broad black band across centre. Red eye surrounded by red eye-ring. Yellow legs. **HABITS AND HABITAT** Common resident of forested mountains above 1,000m. Usually seen in flocks of up to about 30 birds, flying over canopy or perched on tree tops often near oaks, where it feeds on acorns; also eats berries. Call 5–10 notes, 'hu-hooo – hu-hooo – hu-hooo – hu-hooo – hu-hooo'. **SITES** Savegre Valley, Cerro de la Muerte and many mountain areas.

Pale-vented Pigeon ■ *Patagioenas cayennensis* 30cm

DESCRIPTION Fairly large pigeon that at a distance looks bronzy-brown. Closer inspection reveals multicoloured head, overall greyish with glossy bronze-green nape, purplish breast, grey belly and white undertail. Grey rump and tail, black bill, and red legs and eyes. **HABITS AND HABITAT** Common resident on both Caribbean and Pacific Slopes in lowlands to 2,000m. Favours forest edges, secondary growth, river edges and open areas, perching conspicuously on dead tree tops or power lines. Often seen alone or in small groups, and feeds on berries and other fruits, and seeds. Call a soft 'woooo-woop-woop-wooo'. **SITES** Easily seen on roadside wires near La Selva OTS and in many coastal areas.

Red-billed Pigeon ■ *Patagioenas flavirostris* 30cm

DESCRIPTION Fairly large pigeon that is overall purplish-grey with very noticeable cream-white (not red) bill, and small red base area to bill. Back, wings, belly, undertail, rump and tail blue-grey, contrasting with pink-purplish head, neck and breast. Bare skin around eyes is red; orange eyes and red legs. **HABITS AND HABITAT** Fairly common resident in northern half of Costa Rica, particularly the northern Pacific area, in open country, farmland, scattered trees and forest edges to 2,000m. Often alone or in pairs, and rarely in flocks. Perches in tree tops and feeds on berries, seeds and corn. Call a repeated 'woooo-woop-wop-per-woo'. **SITES** Caño Negro area, Arenal and even trees around the city of San José.

White-winged Dove ■ *Zenaida asiatica* 28cm

DESCRIPTION Medium-sized dove that is pale buffy-brown overall. Obvious white wing edge leads into white wing-bar that is most obvious on birds in flight. Tail greyer with white tips bordered with black; in flight wing-tips show black. Small black line on lower cheek, pale blue skin around eyes, orange eyes, dark bill and reddish legs. **HABITS AND HABITAT** Common resident in northern Pacific area and Central Valley down to San José. Favours dry thorn scrub with cactus, especially around farmland to 500m. Has also moved into cities and towns. Perches on wires and forages on the ground for seeds, grain and grit, and often seen in flocks. Call is 'hoo-cook-cawooo'. **SITES** Farm edges, roadside verges and wires. Easy to see in Palo Verde.

Inca Dove ■ *Columbina inca* 20cm

DESCRIPTION Small and long tailed, with upperparts and underparts overall pale greyish completely scaled with black. Scaling smaller on head and neck. Distinct rufous primaries show particularly in flight. Long tail edged with white, thin black bill, red eyes and pink legs. **HABITS AND HABITAT** Common in northern Pacific area and Central Valley, spreading south along Pacific coast in dry, scrubby habitat, farmland, grassland and generally open habitat with scattered trees. Forages on the ground for seeds and grit, often in small groups. Call is 'coo-coop', sounding like 'no-hope'. **SITES** Roadsides, farm edges and lodge grounds within its range.

Common Ground Dove ■ *Columbina passerina* 16cm

DESCRIPTION Small, plump dove with short tail. Mostly pinkish-grey with irregular dark spots and patches on wings. Pronounced scaling on upper breast separates it from similar ground doves. Male has pale bluish-grey head. Bill orangey with dark tip, and pink legs. In flight shows chestnut in wings. **HABITS AND HABITAT** Common resident in Pacific north-west area down through Central Valley. Favours dry, scrubby habitat, farmland, grasses, bushes and open areas to 1,400m. Gathers in flocks around human habitation and forages on ground for seeds, insects and grit. Call a repeated 'cwoop-cwoop'. **SITES** Roadsides and farmland; especially easy to see around Palo Verde.

Ruddy Ground Dove ■ *Columbina talpacoti* 17cm

DESCRIPTION Female very similar to Common Ground Dove (see above), but note plain breast and yellower bill. Male has distinctive bright rufous body with pale blue-grey head. Underparts pale brownish, tailed edged with black and black spots on wings. In flight wings show chestnut and black. Red eyes and flesh-coloured legs. **HABITS AND HABITAT** Fairly common resident in lowland areas to 1,200m. Favours open farmland, grassy fields, secondary growth, bushes and scrub, where it forages on the ground for seeds, insects and grit. Often associates with other doves. Call a soft, repeated 'coowup-coowup-coowup-coowup'. **SITES** Easily seen along roadsides in suitable habitat.

White-tipped Dove ■ *Leptotila verreauxi* 28cm

DESCRIPTION Medium-sized dove, brown-grey on back, wings and tail. Underparts pale pinkish-white with buff chest. Head pale whitish with brown-grey crown, white throat and darker hind-neck. Bare, pale bluish skin around eyes, orange eyes, black bill and reddish legs. **HABITS AND HABITAT** Common resident on Pacific Slope and in Central Valley; less common in Caribbean lowlands to 2,000m. Walks on the ground in preferred habitat of forest floors, secondary growth, farmland and even gardens. Bursts into flight when disturbed. Call a low, mournful 'hoo-woooooo'. **SITES** Easy to see on trails in Carara National Park, La Selva OTS and forest of Palo Verde.

Smooth-billed Ani

■ *Crotophaga ani* 35cm

DESCRIPTION Almost identical to Groove-billed Ani (see opposite), with black, scaly-looking head, neck and breast. Wings, head, back and tail slightly more glossy black; long, slightly fan-shaped tail. Bill differs from Groove-billed Ani's by being even deeper and entirely smooth. Grey-black bill and legs. **HABITS AND HABITAT** Fairly common resident in southern Pacific area only. Almost no overlap with Groove-billed Ani. Favours similar habitat with fields, farmland and marshes. Likes to follow cattle and livestock, where it can be seen in groups of up to 10 chasing and catching insects. Call a long, high squeak, 'wee-eeeeek'. **SITES** Any suitable grassy farmland habitat within southerly range.

Groove-billed Ani

■ *Crotophaga sulcirostris* 30cm

DESCRIPTION Entirely black with scaly-looking head, neck and breast. Wings and tail are slightly more glossy black, and long tail is often partly fanned and hangs down as the bird perches upright. Most distinctive feature is grey-black bill, which has a deeply curved upper mandible and horizontal grooves on lower mandible. Grey legs. **HABITS AND HABITAT** Common and widespread resident to 1,500m. Rare in south with little overlap with Smooth-billed Ani (see opposite). Found in pastures, farmland, marshes and gardens, where it is nearly always in groups up to 10 birds. Very social. Drops from perch and runs along the ground to catch insects. Call a sharp 'tsipjoo-tsipjoo-tsipjoo'. **SITES** Easily seen on fences near fields and marshes, especially at Palo Verde and Caño Negro.

Striped Cuckoo ■ *Tapera naevia* 30cm

DESCRIPTION Upperparts brown streaked with black, and underparts dull white with buff wash on plain unstreaked breast and undertail. Obvious white stripe running over and behind eye, black line through eye and thin black moustachial line. Shaggy rufous crest, long, graduated tail, creamy-yellow bill and grey-brown legs. **HABITS AND HABITAT** Uncommon and increasing resident in open, bushy habitat and farmland with scattered trees to 1,000m, where it is most often located by incessant calling from top of exposed bush or tree. Song a two-note, repeated whistle, 'sweep-peep'. **SITES** Listen for song in suitable lowland habitat. Caño Negro is a good area to see it.

Lesser Ground Cuckoo ■ *Morococcyx erythropygus* 25cm

DESCRIPTION Small and mostly brown above with rich orange-brown underparts. Very distinctive head pattern showing bright yellow eye-ring and bare skin in front of eye (lores), and bright blue skin behind eye, all bordered by black and a white stripe above this, over eye. Bill bright yellow with black upper edge, and flesh-coloured legs. **HABITS AND HABITAT** Fairly common resident in lowlands of northern Pacific area, Guanacaste down to Tárcoles. Skulking bird not easy to see as it spends most of its time on or near the ground. Walks and runs in search of insects, and can remain motionless if detected. Call starts off fast, then descends, 'chuw-chuw – chew-chew – pr-pr-prr-prr-prrr-prrr-prrrr-prrrr-prrrrr-prrrrrr'. **SITES** Guanacaste, for example at Palo Verde, Hacienda Solimar and road to Monteverde.

Squirrel Cuckoo ■ *Piaya cayana* 46cm

DESCRIPTION Large and slender with very long, graduated tail. Mostly bright rufous above, paler on throat and neck; grey breast and belly, black undertail and striking black-and-white underside to tail feathers. Bright red eye surrounded by green-yellow skin, and bright green-yellow bill. **HABITS AND HABITAT** Common and widespread resident in pretty much any habitat to 2,400m. Found in gardens, forest edges and open areas with scattered trees, mangroves and roadside scrub. Hops and runs along branches in search of caterpillars, insects and lizards, then makes short, gliding flight between trees. Variety of calls, but most recognizable is loud, explosive 'kip-weeyoo'. **SITES** May be found anywhere, but regular in Carara and La Selva.

Pacific Screech Owl

■ *Megascops cooperi* 23cm

DESCRIPTION Greyish-brown with distinct dark border to facial disk. Prominent black streaking on pale breast and belly. Rest of bird a combination of brown-and-grey mottling with a few whitish patches on upper wings and back, pale eyebrows and small, indistinct ear-tufts. Pale greenish-grey bill and bright yellow eyes. Very similar to Tropical Screech Owl (see below). **HABITS AND HABITAT** Fairly common resident of northern Pacific area and Caribbean lowlands to 1,000m. Small overlap with Tropical Screech Owl. Favours mixed forests, open areas with scattered trees, mangroves and thickets. Hunts at night for insects, beetles and moths. Call a series of low, rolling notes, 'pu-pu-pu-pu-pu-pu-pu'. **SITES** Guanacaste area, Hacienda Solimar and regular around Tárcoles.

Tropical Screech Owl

■ *Megascops choliba* 23cm

DESCRIPTION Greyish-brown with less distinct border to facial disk than previous species. Prominent black streaking on pale breast and belly. Rest of bird a combination of brown-grey mottling with a few whitish patches on upper wings and back; pale greyish and small, indistinct ear-tufts. Pale greenish-grey bill and bright yellow eyes. Very similar to Pacific Screech Owl (see above). **HABITS AND HABITAT** Fairly common resident of foothills and highlands to 1,500m. Only in Central Valley and on Pacific Slope south from Tárcoles. Favours secondary growth woodland, gardens and coffee plantations. Hunts at night for insects, spiders and even bats. Call a rolling 'prrrrrrr – poo-poo'. **SITES** Lodge gardens including Bougainvillea Hotel and Talari Mountain Lodge.

Bare-shanked Screech Owl

■ *Megascops clarkii* 25cm

DESCRIPTION Rich orangey-buff owl with indistinct orangey facial disk, finely mottled underparts, very small ear-tufts and noticeable white patches on shoulder and upper wing. Rest of bird browny-buff mottled and streaked above. Piercing bright yellow eyes and greyish-green bill; feet fleshy-grey. **HABITS AND HABITAT** Fairly common to uncommon resident of highlands from 900m to timberline. Favours dense montane forest and forest edges; less so but also in suburban gardens and parks. Hunts at night for beetles and other insects, and small rodents. Call a repeated, low 'poo-poo-poop – poo-poo-poop', often delivered from middle to high elevation in tree. **SITES** Possible to find in Savegre Valley and Monteverde.

Mottled Owl

■ *Strix virgata* 34cm

DESCRIPTION Medium-sized, round-headed owl with no ear-tufts. Overall brown-grey, with belly grey-white and heavily streaked black. More mottled than streaked on upper breast; mottled face with greyish eyebrows. Back upper wing spotted buffy-white, and dark banded tail. Dark brown eyes, yellow-green bill and brown-grey legs. **HABITS AND HABITAT** Fairly common and widespread resident to 1,800m, where it prefers open forests and forest edges, gardens and coffee plantations. Roosts low in tangles and hunts at night for large insects, reptiles and small mammals. Call sounds like barking dog, 'wooow-wooow – woo-woo-woo'. **SITES** Many places, but Monteverde, La Selva and Rancho Naturalista are particularly good for seeing it.

Black-and-white Owl ■ *Strix nigrolineata* 38cm

DESCRIPTION Upperparts sooty-black; face and head sooty-black. Underparts, including collar, finely barred black and white, and eyebrows finely speckled black and white. Short tail banded black and white. Orange-brown eyes, orange bill and orange-yellow feet. **HABITS AND HABITAT** Fairly common resident of both Caribbean and Pacific lowlands to 1,500m. Favours lowland wet forests, gardens and mangroves. Strictly nocturnal and rests during day in low thickets often near water. Feeds on large insects and small rodents, but also well known for taking bats in flight as they feed on insects near lights. Call a 'oo-oo-oo-oo-hooo'. **SITES** Around Tárcoles area and occasionally at or near Cerro Lodge.

Spectacled Owl ■ *Pulsatrix perspicillata* 48cm

DESCRIPTION Large, dark brown owl with striking black-and-white face pattern that recalls huge, incomplete white spectacles. Neck and upper breast dark brown, and creamy-white lower chest and belly. Rounded head shows no ear-tufts, and large yellow eyes are distinctive. Creamy-white bill and grey legs. **HABITS AND HABITAT** Fairly common and widespread resident to 1,500m. Favours dense forest and forest edges, as well as trees alongside rivers and streams. Sometimes found roosting in pairs at mid-height, and occasionally active during the day. Hunts insects, lizards and small mammals such as opossums and skunks. Call a deep, vibrating 'woo-woo-woo-woo-woo-woo'. **SITES** Two good spots are Carara NP and La Quinta Country Inn, Sarapiqui.

Costa Rican Pygmy Owl

■ *Glaucidium costaricanum* 15cm

DESCRIPTION Small with dark rufous-brown upperparts with white spotting; underparts white with thick rufous and black streaking. Upper breast dark rufous with white speckling; head also rufous, with small white spots. Two big black spots (eyes) on back of head surrounded by white. White eyebrows, bright yellow eyes and yellow feet. **HABITS AND HABITAT** Rare to local resident of central highland montane forests from 1,200m to timberline. Favours oak forest and edges, where it calls in the early morning and evening from high in canopy, making it difficult to spot. Feeds on small birds, insects and lizards. Call a long series of fast notes, 'hoo-ho-hoo-hoo-hoo-hoo-hoo-hoo-hoo-hoo-hoo'. **SITES** Difficult to see but present in Monteverde and regularly seen from Savegre Mountain Lodge.

Ferruginous Pygmy Owl

■ *Glaucidium brasilianum* 15cm

DESCRIPTION Small with medium rufous-brown upperparts with white spotting; underparts white with narrow rufous and black streaking. Head-crown rufous with white streaks (not spots), Distinctive black-and-white eye-shape markings on back of neck, bright yellow eyes and yellowish legs. **HABITS AND HABITAT** Common resident in lowlands of north-west to 2,200m, where it favours open savannah, secondary growth, thorn scrub, gardens and coffee plantations. Can be seen during the day hunting for small birds, lizards and insects from low perch. Calls incessantly and often mobbed by small birds. Call a continuous 'bip-bip-bip-bip-bip-bip-bip-bip-bip-bip-bip-bip-bip-bip'. **SITES** Mostly Guanacaste, and easy to find at Palo Verde and Hacienda Solimar.

Great Potoo

■ *Nyctibius grandis* 51cm

DESCRIPTION Largest of the potoos; mostly finely mottled grey-buffy-brown with fine barring on underside, odd dark spots on upper chest and back, and large, pale greyish-buff head. Eyes large and brown, but it is normally asleep during the day so they are not visible. Dark bill mostly hidden. **HABITS AND HABITAT** Uncommon resident of wet lowlands on Caribbean Slope and also Osa Peninsula to 800m. Favours dense, mature wet forests, where it perches during the day in canopy. Regularly perches on same perch each day and extremely well camouflaged, looking like part of the tree. Hunts at night for flying beetles and moths. Call an eerie growl, 'wrrrroowl-wrrrroowl'. **SITES** Day roosts often near Arenal Observatory Lodge, La Selva and Caño Negro.

Common Potoo ■ *Nyctibius griseus* 38cm

DESCRIPTION Simple to identify really – it looks just like a broken tree stump! Combination of greyish-brown mottling, streaks and spots, similar to a nightjar's, slender body and huge, bulbous yellow eyes that give off bright eye shine if caught at night in spotlight. Generally only tip of dark bill visible. **HABITS AND HABITAT** Fairly common resident of lowland foothills on Caribbean Slope and southern Pacific area to 1,250m. Favours open forests and woodland, and farmland edges. Perches low down during the day, with cryptic camouflage making it look just like a tree stump or post. Hunts at night for moths and flying insects. Call a strange 'wahooo-oooo-ooo-oo-oo-oo'. **SITES** Good views of it may be had along the roads at night near Caño Negro.

Lesser Nighthawk ■ *Chordeiles acutipennis* 22cm

DESCRIPTION Mostly greyish-buff mottled upperparts; underparts paler grey with fine barring, and long dark wing feathers that reach tail-tip when perched. In flight, male shows white throat and white patch close to end of underwing. Female shows buff throat and buff wing-patch. **HABITS AND HABITAT** Common resident along Pacific coast, and common migrant along both coasts in September–November. Favours lowland scrub and mangroves, where it perches on horizontal branches during the day. At dusk forages over fields, mangroves and rivers, feeding on a variety of insects, sometimes in huge numbers. Mostly silent. **SITES** Many coastal areas, but great numbers can be seen at dusk from Tárcoles River bridge.

Pauraque ■ *Nyctidromus albicollis* 28cm

DESCRIPTION Chunky, medium-sized nightjar with long tail. Upperparts mottled brownish-grey with buff spots on wings; dark throat and white line below throat, and finely barred breast. Male has long tail edged with white, and broad white patches near ends of wings, all obvious when it flies. Female duller, lacking white in tail and wings. Tiny bill and feet are dark. **HABITS AND HABITAT** Common and widespread resident of lowlands to 1,700m. During the day rests on the ground among leaves, where it relies on its cryptic camouflage to avoid detection. Favours open, grassy areas and hunts at dusk, flying up from the ground. Calls 'dweool' or 'dweo-woo'. **SITES** Many sites, but easy to see at La Selva, Arenal and Caño Negro.

White-collared Swift ■ *Streptoprocne zonaris* 22cm

DESCRIPTION Huge swift that is mostly black but in good light shows an obvious unbroken white collar around neck, duller on breast than back of neck. Notched tail and long, swept-back wings. Shape and size like that of small falcon. **HABITS AND HABITAT** Common and widespread resident seen at all elevations. Most often observed in groups of 15 to several hundred birds, which circle and wheel in the air together. Flies very fast and with little effort, hardly flapping its wings. Groups often accompanied by smaller swift species. Feeds on the wing, catching flies, bugs and other insects. Makes bat-like screams and squeals. **SITES** Spectacular and very low overhead on the track above Savegre Mountain Lodge.

Green Hermit ■ *Phaethornis guy* 15cm

DESCRIPTION Fairly large, all-green hummingbird. Long white central tail feathers are diagnostic; rest of tail feathers tipped white, and long, curved bill has pale yellow lower mandible. Male darkish green all over with thin pale line behind bill. Female has two white lines on face, orangey throat and grey underparts. **HABITS AND HABITAT** Common resident of wet montane forests of Central Valley at 500–2,000m. Feeds very low among flowering scrub and bushes, including *Heliconia*, in wet montane forests. Regular route revisiting same patches of flowers. Groups of males gather together in dense understorey at a display ground (lek), where they call repeatedly, uttering a single, high-pitched squeak. **SITES** Easiest to observe on feeders at Monteverde, Rancho Naturalista and Mariposario near San Ramon.

Long-billed Hermit ■ *Phaethornis longirostris* 15cm

DESCRIPTION Overall impression is of a rather large, buffy-brown hummingbird, but closer inspection reveals it to be a bird with more olive-brown upperparts, distinct whitish stripes on the face, plain buffy-grey underparts and long, white-tipped tail with central two tail feathers very long and white. Bill also very long and curved, with dark upper mandible and yellow lower mandible. **HABITS AND HABITAT** Common in lowlands of Pacific and Caribbean Slopes and rarer to 1,000m. Prefers to be near streams in understorey of wet forests. Has a regular circuit of feeding on its favourite flowers, and also eats insects. Males gather at lek and call repeated 'tsweep-p'. **SITES** Most lowland wet forests. La Selva and Carara are good sites in which to see it.

Stripe-throated Hermit ■ *Phaethornis striigularis* 9.5cm

DESCRIPTION Very small, brownish-rufous bird with whitish facial stripes, fairly long, slightly decurved bill, rufous-cinnamon breast, and buffy belly and undertail. No streaking on throat. Graduated tail with buff tips; central tail feathers longer and white. Dark upper mandible and yellow lower mandible. **HABITS AND HABITAT** Common resident throughout lowlands to 1,500m. Feeds very low in forests, forest edges, gardens and semi-open areas. Has a regular circuit to feed on its favourite flowers, some of which it pierces at bases to get to nectar. Males gather at lek close to the ground and give repeated high-pitched, shrill song, 'tseep-tseep-tseep'. **SITES** Arenal Observatory Lodge gardens are a good spot in which to see it.

Violet Sabrewing

■ *Campylopterus hemileucurus* 15cm

DESCRIPTION Large hummingbird with mostly deep violet plumage and obvious white on outer tail feathers, which are often spread. Slightly decurved black bill and small whitish (postocular) spot behind eye. Female dark green above, grey below, with purple throat and white edges to tail. **HABITS AND HABITAT** Common resident in montane wet forests at 1,000–2,400m. Prefers to be near mountain streams, forest edges and banana plantations, and readily comes to hummingbird feeders and flowers. Up to 10 males form leks in thick understorey, where they give a series of high-pitched and variable notes. **SITES** Rancho Naturalista, Savegre Mountain Lodge and Monteverde.

White-necked Jacobin

■ *Florisuga mellivora* 12cm

DESCRIPTION Fairly large with bright royal blue head and upper breast; green back separated from head by narrow, pure white collar; underparts and belly pure white, and tail mostly white with black tips. Bill black, fairly thick and with very slight curve. Female green with speckled upper breast. **HABITS AND HABITAT** Fairly common resident in Caribbean lowlands and Pacific lowlands south from Tárcoles, to 1,000m. Favours forest edges and clearings, and readily comes to hummingbird feeders. Often hovers and perches high in canopy on dead branches. Call a high-pitched 'tseep'. **SITES** Carara NP, Braulio Carrillo and feeders at Mariposario near San Ramon.

Lesser Violetear

■ *Colibri cyanotus* 11cm

DESCRIPTION Medium-sized hummingbird with almost entirely glistening green plumage, and obvious violet ear patches that can be flared in display. Blue-black band on end of tail and thin, very slightly decurved black bill. **HABITS AND HABITAT** Common resident of highlands and mountains mostly above 1,200m, to 3,000m. Found in most montane habitats, including forest edges and clearings, gardens and roadside scrub. Visits wide range of flowers and shrubs; very territorial and aggressive to other hummingbirds. Readily visits hummingbird feeders. Gives incessant clock-like call, 'tsik-tock-tsik-tock-tsik-tock'. **SITES** Common on hummingbird feeders at Savegre Mountain Lodge and Paraiso Quetzales.

Violet-headed Hummingbird ■ *Klais guimeti* 8cm

DESCRIPTION Small, bronzy-green hummingbird with bright violet-blue head and obvious white square (postocular) spot behind eye. Bluish-black tail with pale tips, mostly greyish underparts and shortish, straight black bill. Female same as male, but with green not blue on head, and grey of belly going up to throat. **HABITS AND HABITAT** Fairly common resident in foothills and lowlands of Caribbean Slope to 1,000m and South Pacific Slope to 1,200m. Favours trees and shrubs with small flowers in canopy, as well as forest edges, gardens and secondary growth. Males form leks and also sing from high dead twigs. **SITES** Easy to see on flowers at Arenal Observatory Lodge.

Black-crested Coquette ■ *Lophornis helenae* 7cm

DESCRIPTION Very small and almost insect-like, with intricate plumage. Male has green back and head, pale underparts that are spotted with gold, broad white band on rump and rufous tail. Head has long, wire-like black crest feathers, and black throat and necklace. Short, straight bill bright red with black tip. Female duller than male and lacks all the finery, but still has spotted belly. **HABITS AND HABITAT** Uncommon resident of northern Caribbean foothills and lowlands to 1,200m. Favours canopy, but can come down to feed on small flowers and shrubs in gardens. Perches on dead twigs and flies and hovers with tail cocked. Mostly silent. **SITES** Regular on flowers in Arenal Observatory Lodge gardens, plus Rancho Naturalista.

Green Thorntail ■ *Discosura conversii* 10cm

DESCRIPTION Very small bodied and mostly dark green, but has white band on rump similar to that of a coquette. Male has long, thin, wire-like tail; uppertail blue-black and thigh-patches white. Female lacks long tail and has short, white-tipped tail, white thighs and broad white stripe on face. Bill short, straight and black. **HABITS AND HABITAT** Fairly common to uncommon resident of Caribbean Slope through northern Central Valley to 1,500m. Favours tree tops and forest canopy, where it feeds on flowering trees, occasionally coming lower to feed on small flowers and shrubs. Mostly silent. **SITES** Flowering plants around Braulio Carrillo and feeders at La Paz Waterfall Gardens.

Fiery-throated Hummingbird ■ *Panterpe insignis* 11cm

DESCRIPTION Initial impression is of an all-dark green hummingbird with steel-blue tail and small white (postocular) spot behind eye. When light catches plumage at a certain angle, throat and breast explode into a mass of colour, with red throat surrounded by orange-gold, bright blue sides and chest, and blue crown. Thin, straight bill shows a little pink on underside. **HABITS AND HABITAT** Common resident of high montane forests and scrub, mostly above 2,000m. Can be found in canopy, along forest edges and in roadside flowers, where it is often noisy and frenetic. Call like sound of a very fast, high-pitched squeaky toy, 'zip-zip-zip-zip-zip-zip-zip-zip'. **SITES** Best place to see it is Savegre Valley, but many birds come to feeders at Paraiso Quetzales.

Coppery-headed Emerald ■ *Elvira cupreiceps* 7.5cm

DESCRIPTION Small, with male being green-bronze above and glittering green below, and having a distinct coppery crown and rump. Central tail feathers also coppery, and outer tail feathers and thighs white. Female lacks coppery crown and rump, and has narrow black band on tail and dull white underparts. Thin, slightly curved black bill separates it from similar species. **HABITS AND HABITAT** Endemic to Costa Rica. Fairly common resident of North Caribbean and Pacific Slopes at 400–1,500m. Favours wet montane forests and forest edges, and feeds at all levels from canopy to low shrubs. Males form middle-elevation leks and give series of high-pitched notes. **SITES** Mariposario near San Ramon and La Paz Waterfall Gardens.

Crowned Woodnymph ■ *Thalurania colombica* 10cm

DESCRIPTION Male looks dark but certain light reveals crown, shoulders and belly to be a dazzling purple, with rest of upperparts being green, and throat and upper breast dazzling green. Tail deeply forked and blue-black. Thin, almost straight bill, black above and mostly red below. Female green with white throat and undertail; breast greyish-green and rounded tail with white tips. **HABITS AND HABITAT** Common resident of Caribbean Slope and southern Pacific lowlands to 1,200m. Favours lower elevation forests, forest edges and savannah. Call an incessant 'tsip-tsip-tsip'. **SITES** Easily seen at Mariposario near San Ramon.

Cinnamon Hummingbird ■ *Amazilia rutila* 10cm

DESCRIPTION Medium-sized hummingbird with completely cinnamon underparts and rufous tail. Upperparts bronzy-green and bill bright red. Female duller than male, with all-black bill. **HABITS AND HABITAT** Fairly common resident of north-west lowlands, including Guanacaste and Nicoya Peninsula to 1,000m. Favours savannah and dry deciduous forests, forest edges, and scrubby bushes and gardens, where it feeds on flowering trees and shrubs. Males sing from dense scrub and can be aggressive towards other hummingbirds. Call a sharp 'tsip'. **SITES** Guanacaste and Palo Verde areas; feeds in gardens of Cerro Lodge.

Rufous-tailed Hummingbird ■ *Amazilia tzacatl* 10cm

DESCRIPTION Medium-sized hummingbird with green upperparts, glistening green throat and upper breast contrasting with greyish belly. Bright rufous tail and uppertail, and bright red, slightly decurved bill with black tip. Female a duller version of male, with less glistening green. **HABITS AND HABITAT** Widespread and common resident over much of Costa Rica from sea level to 2,000m. The most abundant hummingbird, found in a wide variety of habitats, but preferring open deforested areas, parks, gardens, scrub, coffee plantations and secondary growth. Very aggressive and territorial, chasing off all other hummingbirds. Call a low, repeated 'tzip-tzip-tzip-tzsip'. **SITES** Likely to be found in any flowering gardens and on most hummingbird feeders.

Steely-vented Hummingbird ■ *Amazilia saucerottei* 9cm

DESCRIPTION Medium-sized hummingbird showing mostly bronzy-green upperparts; glittering green underparts, and undertail steel-blue with feathers edged buff. Coppery rump and uppertail and dark steel-blue tail. Small white thigh-tufts and slim, straight bill mostly black with pinkish lower mandible. **HABITS AND HABITAT** Fairly common resident of lowlands of North Caribbean Slope, across Central Valley to northern half of Pacific Slope, ranging to 1,200m. Favours open dry savannah, scattered trees, secondary growth, coffee plantations and forest edges. Feeds on flowering trees and shrubs and often aggressive in defending its patch of flowers. Call a typical high-pitched 'tsip'. **SITES** Regular around Carara NP, Cerro Lodge and Villa Lapas.

Snowcap ■ *Microchera albocoronata* 6.5cm

DESCRIPTION Tiny hummingbird with male an unmistakable red-wine colour, topped off with brilliant snow-white crown. Female much more difficult to identify, but note tiny size, bronzy-green upperparts, dull grey-white underparts, whitish tip to tail and small white (postocular) spot behind eye. Bill short, thin and straight. **HABITS AND HABITAT** Uncommon and localized resident along Caribbean Slope at 300–1,200m. Prefers to feed on small flowers in canopy along forest edges, but ventures down to flowering scrubs and even feeders. Often chased off by larger hummingbirds. Call a very thin 'tsew'. **SITES** Celeste Mountain Lodge, Rancho Naturalista and disused butterfly garden by Braulio Carrillo.

Grey-tailed Mountaingem ■ *Lampornis cinereicauda* 10cm

DESCRIPTION Medium-sized hummingbird. Male has green-blue upperparts, greenish breast, grey belly and clean white throat. White line sweeps back behind eye, and seen from a certain angle the crown flashes a brilliant aquamarine blue. Tail dull greyish. Female has completely cinnamon underparts, and white line behind eye is bordered by black. **HABITS AND HABITAT** Fairly common resident in highlands of south-central Costa Rica, generally above 1,800m to timberline. Feeds on flowers of bromeliads and epiphytes. Usually silent. **SITES** Easy to see on feeders at Savegre Mountain Lodge.

Green-crowned Brilliant ■ *Heliodoxa jacula* 13cm

DESCRIPTION Large and robust hummingbird. Male almost entirely sparkling green, with violet spot in centre of throat. Small white (postocular) spot behind eye and white thigh-tufts. Long and deeply forked, blue-black tail. Female similar to male but with greyish underparts speckled with green spots, and white line on face. Bill black, thick and stout. **HABITS AND HABITAT** Fairly common resident of central mountain ranges at 700–2,000m. Favours wet highland forests, where it feeds on epiphytes and understorey flowers. Perches to feed and does not hover. Call a loud 'tchew-tchew-tchew-tchew'. **SITES** Can be expected at Paraiso Quetzales feeders and Mariposario near San Ramon.

Talamanca Hummingbird ■ *Eugenes spectabilis* 13.5cm

DESCRIPTION Large and robust hummingbird. Male upperparts and breast dark bronzy-green, greyish belly and blackish sides of head. Crown a dazzling purple seen in the right light, and throat an equally dazzling blue. Shows obvious white (postocular) spot behind eye, and long straight, sturdy bill is black. Female green with dull grey underparts. **HABITS AND HABITAT** Common resident of highlands of southern and central Costa Rica from 1,500m to timberline. Favours montane oak forests and forest edges, where it feeds on flowers, epiphytes and thistles along a regular circuit. Call a single or rattling 'prrrt' or 'prrrrrrrrrt'. **SITES** Feeders at Paraiso Quetzales and Savegre Mountain Lodge.

Volcano Hummingbird ■ *Selasphorus flammula* 7.5cm

DESCRIPTION Very small and insect-like hummingbird, with bronzy-green upperparts and mostly white underparts with dusky flanks. White from upper breast forms collar. White (postocular) spot and thin, straight bill. Tail rufous with central two feathers bronze-green. Female has speckled pale throat. Depending on range, the three races show different-coloured gorgets on males, from red and purple, to greyish-green. **HABITS AND HABITAT** Common resident of highlands in central and southern mountains from 1,800m to highest peaks. Favours paramo, open areas, scrubby hillsides and elfin forest. Feeds on nectar through hole in flower made by bees and flowerpiercers. Mostly silent. **SITES** Top of any accessible volcano, Cerro de la Muerte and Savegre Mountain Lodge.

Scintillant Hummingbird ■ *Selasphorus scintilla* 7.4cm

DESCRIPTION Fractionally smaller than Volcano Hummingbird (see above), and can be distinguished from that species by dazzling orange gorget, rufous-cinnamon on flanks and more rufous on tail, with less obvious pale (not dark) brown-green central tail feathers. Tiny, thin bill, white collar and (postocular) spot similar to Volcano Hummingbird's. Female has finer speckling on throat than male. **HABITS AND HABITAT** Scarce and local resident of highlands and mountains of central Costa Rica at 900–2,400m. Feeds on low flowering bushes and shrubs along forest edges, open areas and even gardens. Mostly silent but in flight wings give insect-like buzz. **SITES** One or two come to feeders at Savegre Mountain Lodge.

Resplendent Quetzal

■ *Pharomachrus mocinno* 36cm, plus 70cm streamers

DESCRIPTION Often described as one of the most beautiful birds in the world, the male of this large trogon displays four very long st[illegible] in iridescent green, and extended wing-coverts that hug the sides of the brilliant red breast. Male also has bright iridescent green upperparts, head and upper breast, crest with gold fringe that covers part of bill, long white undertail and yellow bill. Female duller than male, with no crest or streamers, and grey breast and head. **HABITS AND HABITAT** Fairly common resident of epiphyte-laden mountain forests at 1,200–3,000m. Nests in hole in tree and feeds on avocados and other fruits. Call a deep 'klook-kloo – klook-kloo'. **SITES** Savegre Mountain Lodge, Paraiso Quetzales and Trogon Lodge.

Slaty-tailed Trogon

■ *Trogon massena* 30cm

DESCRIPTION Large trogon with bright red belly. Male has glossy green head, breast, uppertail and back, and black face; wing-coverts finely barred grey, and undertail clean unbarred slaty-grey. Red eye-ring, orange eyes and orange-red bill. Female mostly grey with red lower belly, and bill with dark upper mandible and orange lower mandible. Grey feet. **HABITS AND HABITAT** Fairly common resident of wet lowland forests on Caribbean Slope and South Pacific Slope to 1,200m. Favours wet, humid forests and forest edges, where it often sits high and feeds on fruits, insects and lizards. Call a repeated, low 'haah-haah-haah-haah-haah-haah'. **SITES** Arenal Observatory Lodge and Carara area.

Black-headed Trogon ▪ *Trogon melanocephalus* 28cm

DESCRIPTION One of three yellow-bellied trogons found in Costa Rica, but the only one with pale blue eye-ring and blue-grey bill. Male has dusky-black head and upper breast, glossy green back, deep blue rump and bluish-green uppertail with black tip. Three broad white bands on undertail. Female similar to male, but head and upperparts greyish, and bill with dark upper mandible. **HABITS AND HABITAT** Common resident of lowlands of northern Costa Rica to 800m. Favours dry forests and forest edges, and secondary growth in savannah. Feeds on fruits and insects. Call an accelerating 'chew-chew-chu-chu-chu-chu-ch-ch-ch-ch-ch'. **SITES** Fairly easy to see at Palo Verde and in Guanacaste area.

Baird's Trogon ▪ *Trogon bairdii* 28cm

DESCRIPTION Back glossy green-blue, wings dark grey-black with pale grey edges to primaries. Outer tail glossy green-blue, undertail mostly white, underparts from lower chest to undertail rich red, upper breast and head violet-blue and face blackish. Distinctive blue eye-ring, creamy-blue bill and grey legs. Female completely sooty-grey, with red lower belly and undertail. **HABITS AND HABITAT** Fairly common resident of lowlands in southern Pacific area to 1,200m. Prefers to sit high up in canopy of tall, mature rainforests, where it can be difficult to spot. Will come lower to feed on fruiting trees. Song a slow chuckle accelerating before an abrupt stop, 'chup-chup-chup-chup-chup-chp-chp-chpchpchpchpchp-chuperr'. **SITES** Carara NP and Bosque del Rio Tigre.

Gartered Trogon

■ *Trogon caligatus* 24cm

DESCRIPTION Yellow-bellied trogon with bright yellow eye-ring on male. Female with incomplete white eye-ring. Male head and upper breast violet-blue; black face and glossy green back. Grey-looking wings, grey bill and feet, and undertail barred black and white with broad white tips. Female's head and upperparts greyish. **HABITS AND HABITAT** Common and widespread resident of both Caribbean and Pacific lowlands to 1,200m. Favours both dry and humid forest edges, open areas, secondary growth and gardens, where it feeds on fruits and insects. Call a repeated 'kew-kew-kew-kew-kew-kw-kw-kw-kw-kw-kw-k-k'. **SITES** Carara NP and Bosque del Rio Tigre.

Black-throated Trogon

■ *Trogon rufus* 23.5cm

DESCRIPTION The third yellow-bellied trogon of Costa Rica and the smallest. Male has pale blue eye-ring and yellow bill. Iridescent green head, upper breast and upperparts. Blackish face and grey-looking wings. Undertail finely barred black and white, with broad white tips. Female's head and upperparts brown, incomplete white eye-ring and black edge to upper mandible. **HABITS AND HABITAT** Common resident of wet lowlands on Caribbean and South Pacific Slopes to 1,200m. Favours mature, dark and shady wet forests, where it feeds on small fruits, insects and caterpillars. Raises tail when calling. Call a mournful, slow 2–5 notes, 'croow-croow-croow-croow-croow'. **SITES** Reliable at La Selva OTS and Bosque del Rio Tigre.

Collared Trogon ■ *Trogon collaris* 25cm

DESCRIPTION Medium-sized, red-bellied trogon with obvious white breast-band. Male has glossy metallic green head, upper breast and upperparts, black face and finely barred black-and-white undertail with white tips. No obvious eye-ring and yellow bill. Female has white breast-band, but upperparts, head and upper breast are brown; pale broken eye-ring. Bill with dark tip. **HABITS AND HABITAT** Fairly common resident of central and southern foothills and highlands at 600–1,500m. Favours wet montane forests and forest edges, as well as gardens and secondary growth. Feeds on beetles, crickets and caterpillars. Call is 2–4 clear notes, 'caow-caow-caow'. **SITES** Monteverde, El Toucanet Lodge and Savegre Mountain Lodge.

American Pygmy Kingfisher ■ *Chloroceryle aenea* 13cm

DESCRIPTION Smallest of the Neotropic kingfishers and because of its size is easily overlooked. Typical kingfisher shape with dark metallic green head, upperparts and tail, small, light buff-white spots on wings, rufous-orange from throat, through breast and belly, white undertail and mostly black bill. Female has narrow green breast-band. **HABITS AND HABITAT** Fairly common resident along small woodland streams, ponds, rivers and edges of mangroves where vegetation is dense. Often perches low, close to water, where it can be very approachable, often pumping its tail and bobbing its head before plunging in for small fish and aquatic insects. Call a high-pitched whistle. **SITES** Probably best seen on boat trips at Rio Tarcoles, Caño Negro and Tortuguero.

Green Kingfisher ■ *Chlorocerlye americana* 18cm

DESCRIPTION Small kingfisher; male has metallic green head, upperparts and tail, and distinctive white spots on wings and white on tail edges. Broad white collar and throat, and rufous breast-band. Underparts white with spots. Female has two green breast bands and no rufous. Bill sturdy and black, and feet greyish. **HABITS AND HABITAT** Common and widespread resident of lowlands to 1,200m. Found in pretty much any lowland wet habitat, including streams, rivers, ponds and mangroves. Typically perches low on twigs, rocks or fences, from where it plunge dives after small fish. Call sounds like two pebbles being tapped together, 'tt-tt'. **SITES** Easy to see on Rio Tárcoles, Caño Negro and Tortuguero.

Amazon Kingfisher ■ *Chloroceryle amazona* 28cm

DESCRIPTION Similar in looks to Green Kingfisher (see above), but much larger and has no white spots in wing and no white in tail. Upperparts and head metallic green, and bold white collar and throat. Underparts of male have broad rufous breast-band, and white belly with green spotting. Female has no rufous and an incomplete green breast-band. Both sexes have a shaggy crest. Long and stout black bill. **HABITS AND HABITAT** Fairly common resident of lowlands to 1,200m. Favours fast-flowing rivers, ponds and mangroves. Hunts for fish by plunge diving and hovering. Call a harsh rattle, 'tzik-tzik-tzik'. **SITES** Easy to see on Rio Tárcoles, Caño Negro and Tortuguero.

Ringed Kingfisher ■ *Megaceryle torquata* 41cm

DESCRIPTION Very large kingfisher. Male has blue-grey upperparts and head, shaggy crest, broad white collar and throat, and rich rufous underparts. Longish tail chequered black and white; small white spot in front of eye. Female similar to male, but chest is blue-grey and only belly is rufous. Large, stout bill is dark grey. Legs brownish-grey. **HABITS AND HABITAT** Fairly common resident in lowlands over much of Costa Rica to 1,200m. Favours rivers, lakes, lagoons, estuaries and mangroves. Generally sits low, but can perch high in trees overhanging water, where it plunge dives for fish. Call a loud 'krrrik' and machine gun-like rattle. **SITES** Easy to see during boat trips on Rio Tárcoles, Caño Negro and Tortuguero.

Lesson's Motmot ■ *Momotus lessonii* 40cm

DESCRIPTION Large motmot with green upperparts shading to blue on tail; underparts olive-green, sometimes with rusty tinge on breast. Black spot in centre of breast. Head pattern includes black mask through eye, slightly blue throat and bright blue stripe on sides of dark crown. Long tail has two bluish rackets at end. Blackish bill and feet, and bright red eyes. **HABITS AND HABITAT** Fairly common resident of Central Valley and Pacific lowlands to 2,000m. Favours shady forests and gardens, as well as secondary growth and forest edges. Perches low and wags tail from side to side. Feeds on insects, lizards and fruits. Call a hollow 'boop-boop – boop-boop'. **SITES** Many places, including Talari Mountain Lodge and La Quinta Country Inn.

Rufous Motmot

■ *Baryphthengus martii* 46cm

DESCRIPTION Largest of the Costa Rican motmots. Green back and wings, and undertail with long blue tail with two small rackets at the end. Rest of underparts and head bright rufous; small black spot in centre of chest, and black mask through eye. Bill rather heavy looking and black, red eyes and grey feet. **HABITS AND HABITAT** Fairly common resident of Caribbean lowlands to 1,000m. Favours mature wet forests and shady understorey. Tends to perch low and feeds on insects, lizards and even land crabs. Call a loud, echoing 'boo-book – boo-book'. **SITES** Regular at La Selva OTS and Tirimbina Lodge.

Broad-billed Motmot

■ *Electron platyrhynchum* 30cm

DESCRIPTION Initially looks like small Rufous Motmot (see above), but rufous underparts only extend to upper breast (not belly). Green belly and undertail, blue throat-patch and black spot in centre of chest. Long tail bluish-green with two bluish rackets at end. Small black mask goes through eye. Blackish bill and grey feet. **HABITS AND HABITAT** Common resident of Caribbean lowlands and foothills to 1,000m. Favours humid and wet forests, forest edges and semi-open areas, where it sits mostly low and in a shady area. Feeds on fruits, small insects, butterflies, dragonflies and ants. Call a grating, harsh 'kroaw'. **SITES** Fairly easy to see around Arenal Observatory Lodge and La Selva OTS.

Turquoise-browed Motmot ■ *Eumomota superciliosa* 34cm

DESCRIPTION This beautiful little motmot displays a multitude of colours. Rufous back and undertail, green wings with turquoise flight feathers, turquoise tail with rackets tipped with black at the end of long, wire-like feather shafts, green-olive breast, face and neck, black mask through eye, rufous patch behind eye, striking aqua-blue crown and blue throat with black centre patch. Bill slender and black. **HABITS AND HABITAT** Common resident of north-west Pacific lowlands and Guanacaste to 900m. Favours dry, scrubby savannah, secondary growth and thickets, as well as forest edges and mangroves. Perches silently with tail wagging, and fly catches for insects. Call a rasping 'kworrk-kworrk'. **SITES** Guanacaste area, Palo Verde and Santa Rosa.

Rufous-tailed Jacamar ■ *Galbula ruficauda* 23cm

DESCRIPTION Upperparts, head, tail and breast-band a dazzling iridescent green; underparts and undertail rich rufous. Male has white throat; buff-cream in female. Black bill very distinct, being very long, straight and tapering to a sharp point. **HABITS AND HABITAT** Fairly common resident in lowlands of Caribbean Slope and South Pacific Slope to 1,200m. Favours shady forest understorey close to streams, from where it swoops out to catch insects such as butterflies, dragonflies and bees. Always perches low on a horizontal branch or fence wire, and sits with bill pointing slightly up. Call a shrill 'sweeeep-sweeep'; also long, descending trill. **SITES** Carara NP and Arenal Observatory Lodge, La Esquinas and Rancho Naturalista.

Pied Puffbird

Notharchus tectus 15cm

DESCRIPTION Predominately black and white, dumpy and round in shape. Upperparts mostly black with white wing-patch and small white spots on black crown. Thin white line above eye, white throat and underparts, and black breast-band. Shortish tail barred black and white underneath, grey-black feet and sturdy, hook-tipped black bill. **HABITS AND HABITAT** Uncommon and decreasing resident of Caribbean lowlands to 300m. Favours canopy of forest edges, tall secondary growth and large lone trees. Perches high up and hard to spot if not calling. Feeds on insects such as dragonflies, bees and wasps. Call a very high-pitched 'pewee-pewee' and various whistling buzzy notes. **SITES** Regular at La Selva OTS.

White-whiskered Puffbird

Malacoptila panamensis 19cm

DESCRIPTION Dumpy, round-shaped bird, mostly brown-grey. Male has warm brown upperparts, head and chest, and pale greyish lower belly. Chest and belly lightly streaked. Female a colder grey-brown with heavier streaking on pale grey underparts. Long, scruffy white whiskers surround dark grey bill, and eyes are deep red. **HABITS AND HABITAT** Fairly common resident of south, central Caribbean and Pacific lowlands to 1,200m. Favours wet lowland forests and shady undergrowth, where it can sit motionless and undetected for long periods. Feeds on insects, small frogs and lizards, and often accompanies mixed flocks of birds and ant swarms. Call a long, drawn-out 'psseeeeeeeeeeeeew'. **SITES** Carara NP is a great spot in which to see it.

Red-headed Barbet ■ *Eubucco bourcierii* 15cm

DESCRIPTION Both sexes have green back, wings and tail, and lemon belly with greenish streaks on lower belly and undertail. Male has bright red head, throat and upper breast, dark mask around eye and bill, thin white neck-collar, yellow-cream bill and red eyes. Female's head has pale blue face, blue nape and thin orange collar running into breast-band, black forehead and pale yellow bill. Green-grey feet. **HABITS AND HABITAT** Uncommon resident at middle elevation of central Costa Rica at 400–1,800m. Favours wet rainforests, where it forages high in canopy and among tangles for insects, spiders and fruits. Mostly silent, but male's song is a dove-like purr, 'prrrrrrrrrrrrrrrrrrrrrr'. **SITES** Can be seen at Tapantí NP and around La Paz Waterfall Gardens.

Prong-billed Barbet

■ *Semnornis frantzii* 18cm

DESCRIPTION Overall a dirty mustard colour with warmer gold around the face and head. Undertail and flanks paler grey, gold belly, and black mask around eye and bill. Male has thin black ponytail on back of nape. Pale silvery-blue bill with two unusual little prongs on lower mandible near tip. Red eyelids and olive-grey feet. **HABITS AND HABITAT** Fairly common resident of Central Highlands on both slopes at 700–1,400m. Favours moss-covered wet forests, edges and gardens, where it feeds mostly on fruits. Groups and pairs produce a far-carrying duet, 'kwaaa-kwaaa-kwaaa-kwaaa'. **SITES** Tapantí NP and around La Paz Waterfall Gardens.

Blue-throated Toucanet ■ *Aulacorhynchus caeruleogularis* 30cm

DESCRIPTION Smallest of Costa Rican toucans. Entire body mostly bright green with bronze tinge to crown and nape. Undertail chestnut, and face and throat bright blue. Bright white line surrounds bill base, which has a black lower mandible, and mostly yellow upper mandible with a deep red patch at base. Dark eyes and blue-grey feet. **HABITS AND HABITAT** Fairly common resident from middle to high elevation throughout central to southern Costa Rica at 800–2,500m. Found mostly in montane wet forests and forest edges, tall secondary growth and wooded gardens. Feeds mostly on fruits. Call sounds like small barking dog, 'yap-yap-yap-yap'. **SITES** Monteverde, Savegre Valley and Paraiso Quetzales.

Collared Aracari ■ *Pteroglossus torquatus* 41cm

DESCRIPTION Head, neck, throat and back glossy black, narrow chestnut collar, long, graduated black tail and bright red rump. Underparts mostly yellow with black spot in middle of chest, black band with red edges across belly and chestnut thighs. Thin white line separates face from bill. Lower mandible black, upper mandible shading dark through pale yellow to reddish with saw like-pattern. Yellow eyes and greenish feet. **HABITS AND HABITAT** Common resident in lowlands of Caribbean Slope and North Pacific Slope to 1,200m. Usually seen in small flocks of 6–10 birds in wet forests, forest edges and secondary growth, feeding on fruits. Call a very high-pitched, sharp 'tzzeep'. **SITES** La Selva OTS, La Quinta Country Inn and Braulio Carrillo.

Fiery-billed Aracari ■ *Pteroglossus frantzii* 43cm

DESCRIPTION Similar in appearance to the more common Collared Aracari (see opposite). Black head separated from back by chestnut line, yellow breast with black centre spot and red underparts. Eyes an obvious white; key identification feature is bill, which has upper mandible that is fiery orange near tip, shading to bright yellow-green near base. Lower mandible black; obvious white line where bill joins face. **HABITS AND HABITAT** Resident of South Pacific Slope to 1,500m. Often in high canopy of humid forests, clearings, and even secondary roadside scrub and gardens. Call a high-pitched, loud, sharp 'kseek' or 'keeseek'; often with two notes. **SITES** Carara National Park, south to Osa Peninsular, and visits feeders at Cerro Lodge.

Keel-billed Toucan

■ *Ramphastos sulfuratus* 47cm

DESCRIPTION Large toucan that is mostly black with bright yellow face, throat and upper breast. Red undertail, white uppertail-coverts and black tail. Striking bill mostly lime green, with long orange patch through upper mandible, sky-blue patch on lower mandible and deep red tip. Bluish legs. **HABITS AND HABITAT** Common resident of Caribbean lowlands and North Pacific Slope to 1,300m. Favours forested lowlands and foothills, plus pastures with tall trees and secondary growth. Travels in small groups and feeds mostly on fruits, including berries. Call sounds like several frogs, 'creek-creek – creek-cruk – creek-cruk – creek-creek'. **SITES** Many places in Arenal area, Carara NP, La Selva OTS and Monteverde.

Yellow-throated Toucan ■ *Ramphastos ambiguus* 56cm

DESCRIPTION Largest Costa Rican toucan, which is overall mostly black, with bright yellow face, throat and upper breast. Red undertail and white rump. Small red-and-white band separates yellow breast from black belly. Dark eye surrounded by green skin, and two-tone bill mostly dark purple-black with top half of upper mandible yellow. Legs bluish. **HABITS AND HABITAT** Common resident of lowlands and foothills on Caribbean Slope and South Pacific Slope to 1,800m. Small flocks move through forest edges and semi-open areas with tall trees, where they feed on fruits and occasional insects and lizards. Call a 'kireek-tok – kireek-tok – kireek-tok'. **SITES** Carara NP, Las Esquinas, Bosque del Rio Tigre and La Selva OTS.

Olivaceous Piculet

■ *Picumnus olivaceus* 9.5cm

DESCRIPTION Tiny woodpecker. Upperparts brown-olive, wings edged with olive-green and underparts buffy-brown tinged yellowish. Very short tail is black with outer and central feathers creamy-white. Throat white, face buff and rest of upper head black with distinct small white spots on crown. Male has orange on forecrown. Small, sharp black bill and grey feet. **HABITS AND HABITAT** Fairly common resident on South Pacific Slope to 1,400m; rare on North Caribbean Slope. Favours forest edges, tall scrub and gardens, usually keeping low or hanging on small branches or dead trees in search of small insects. Call a high-pitched trill, 'psttttttttttttttttttttt'. **SITES** Las Esquinas and Bosque del Rio Tigre.

Acorn Woodpecker

Melanerpes formicivorus 22cm

DESCRIPTION Striking black-and-white woodpecker. Upperparts mostly glossy black with blue sheen. White rump, belly, forehead and throat. Glossy black mask around bright white eye joins with black of nape and breast-band. Black streaks from breast to flanks. Obvious white patch in wings very evident in flight. Male has red crown; female has black forecrown and red at rear of crown. **HABITS AND HABITAT** Fairly common resident of central southern highlands mostly above 1,500m. Favours montane oak forests where its primary food is acorns, which are stored in crevices and holes in dead trees. Very communal and found in groups. Call a harsh 'rrraaaak' and 'raaker-raaker-raaker'. **SITES** Easy to see at Savegre Mountain Lodge.

Black-cheeked Woodpecker

Melanerpes pucherani 18cm

DESCRIPTION Upperparts including wings and tail are black, and either barred or spotted with white. Underparts mostly grey, with breast and belly finely barred with black and a red centre to belly. Rump white; male has red crown and nape, while female only has red on nape. Forehead golden-yellow, and black mask running through and behind eye down to back. Dark bill and grey legs. **HABITS AND HABITAT** Common resident of Caribbean lowlands and foothills to 900m. Can be found in pretty much any habitat, including gardens and forest edges. Feeds on wide variety of insects, fruits and nectar, and regularly comes to birdfeeders. Call a rattling 'chiirrrrupp-chiirrrrupp'. **SITES** Birdfeeders at La Quinta Country Inn and Arenal Observatory Lodge.

Hoffmann's Woodpecker

■ *Melanerpes hoffmannii* 18cm

DESCRIPTION On initial views resembles an overall grey-looking woodpecker, but closer inspection reveals boldly barred back and wings. Underparts pale grey with dark barring on rear flanks and yellow in centre of belly; plain pale grey face with beady dark eye. Male has red crown with white forehead and yellow nape; female has completely grey head with yellow only on nape. Dark grey bill and legs. **HABITS AND HABITAT** Common resident of northern Pacific and northern Caribbean areas, and Central Valley, to 2,100m. Favours open areas with scattered trees, secondary growth, parks and gardens. Feeds on variety of food, including insects, grubs, fruits and nectar. Call a rattling 'prrr-rr – prrr-rr – prrr-rr'. **SITES** Feeders at Cerro Lodge and many hotel gardens.

Golden-olive Woodpecker ■ *Colaptes rubiginosus* 20cm

DESCRIPTION Upperparts plain bright olive-green, including tail. Underparts from throat to undertail pale with dark olive-brown barring, and chin speckled. Face has large creamy cheek-patch that on male is surrounded by crimson line and crimson nape; female has dark line below and above cheek-patch and crimson only on nape. Slaty-grey crown, black bill, grey legs and deep red eyes. **HABITS AND HABITAT** Fairly common resident of middle-elevation Caribbean and Pacific Slopes at 750–2,100m. Generally found high up in forest edges, shady secondary growth, gardens and coffee plantations. Feeds on insects such as ants and termites. Call a fast, rattling 'dee-dee-dee-dee'. **SITES** Regular all around Arenal area.

Lineated Woodpecker

■ *Dryocopus lineatus* 33cm

DESCRIPTION Large woodpecker with thin neck and distinctive large, crested head. Upperparts, wings, tail and throat slaty-black, and underparts pale grey with smudgy dark barring. Two obvious white stripes from shoulder to middle of black-and-white stripe running from sides of neck, across face to bill. Black line through eye, red below white line, and bright crimson crown and pointed crest. Yellow eyes. **HABITS AND HABITAT** Fairly common resident in lowlands and foothills over much of Costa Rica to 1,200m. Favours dry forests and forest edges, secondary growth and open areas, where it pries open bark in search of grubs and ants. Call a rattling laugh, 'wic-wic-wic-wic-wic-wic-wic'. **SITES** Many areas including La Selva OTS.

Pale-billed Woodpecker ■ *Campephilus guatemalensis* 33cm

DESCRIPTION Similar to Lineated Woodpecker (see above), except that white lines on back join together forming a 'V'; most notably, whole of head is completely crimson and bill is creamy-white. Underparts more heavily barred than in Lineated. Eye yellowish and feet greyish-blue. Female differs from male only by black on front of crest and black throat. **HABITS AND HABITAT** Fairly common resident of lowlands and foothills over much of Costa Rica to 1,400m. Favours more forested areas than previous species, but also found along forest edges and in secondary scrub. Feeds mainly on grubs extracted from decaying trees. Call a sharp 'keew-keew-keew'; distinctive loud double knock as it hammers on a tree trunk. **SITES** Easy to see in Carara NP.

Northern Crested Caracara

Caracara cheriway 60cm

DESCRIPTION Wings, back, under-belly and thighs dark brown, undertail white, and pale breast of adult finely barred with black. Face and neck pale creamy-white with spots and broken barring. Head has dark brown cap with shaggy crest at rear. Bright orange facial skin between eyes and bill. Pale brown eyes and ivory-white bill. Long, bare legs and yellow feet. In flight shows large white outer wing-patch. **HABITS AND HABITAT** Common resident of northern lowlands and southern Pacific lowlands to 1,000m. Favours open country and farmland, and often seen along roadsides, where it frequently feeds with vultures on carrion. Call like a wooden rattle; also a high-pitched squeal. **SITES** Easy to see well on boat trips on Rio Tárcoles.

Yellow-headed Caracara

Milvago chimachima 41cm

DESCRIPTION Upperparts medium brown; underparts, head, neck and thighs creamy-white. Narrow dark line extends behind eye. Yellow skin between eye and bill, which is pale blue. Tail looks whitish with brown bands and blackish terminal band. Bluish legs. Juvenile is browner than adults and has very streaky underparts. **HABITS AND HABITAT** Fairly common resident along Pacific lowlands and northern Caribbean area to 1,500m. Favours open country, savannah, farmland and scrub, where it can perch on tall trees or on the ground. Feeds on carrion, small mammals, lizards and insects. Call a scream-like 'reeeeeeaah'. **SITES** Easily seen from boats on Rio Tárcoles and Caño Negro.

Laughing Falcon ■ *Herpetotheres cachinnans* 52cm

DESCRIPTION Stocky and large headed with longish, rounded tail. Back and wings dark brown, and tail blackish with buff banding. Underparts and broad collar creamy-white, as is rest of head except wide black mask through and behind eye. Dark eyes and yellowish legs and feet. **HABITS AND HABITAT** Fairly common resident in lowlands of both Caribbean and Pacific Slopes to 1,600m. Favours open habitat, forest edges and savannah with scattered trees, on which it often perches high up on open branch. Feeds mostly on snakes and lizards, and often seen alone. Laughing call from which it got its name sounds like a barking dog, 'gua-co – gua-co – gua-co – gua-co'. **SITES** Guanacaste area including Palo Verde.

American Kestrel ■ *Falco sparverius* 26cm

DESCRIPTION Adult male has rich rufous back with large black spots, blue-grey wings and crown, striking black-and-white face pattern, black flight feathers, and long rufous tail with black subterminal band. Buffy breast and white underparts with a few black spots along flanks. Female all brown with no blue-grey, and shows streaked underparts. Brown eyes and yellow feet. **HABITS AND HABITAT** Uncommon migrant in September–April, and winter resident mostly in Pacific lowlands and Central Valley to 1,800m. Favours open farmland and fields, where it often perches on wires and fences. Hovers in search of food such as insects, mice and lizards. Shrill call, 'klie-klie-klie-klie'. **SITES** Palo Verde is a regular spot in which to see it.

Scarlet Macaw ■ *Ara macao* 84cm

DESCRIPTION Large and fantastically bright red macaw with golden-yellow on wings, deep blue flight feathers, blue rump and undertail, large white patch of bare skin on face, and very long, bright red tail with blue tip. Huge hooked beak has pale cream upper mandible and black lower mandible. White eyes and grey feet. **HABITS AND HABITAT** Very localized resident in coastal lowlands, with most pairs restricted to Carara NP, Tárcoles and Osa Peninsula, and just a few in Palo Verde. Favours tall trees in forests and near mangroves, where it feeds on fruits, nuts and various palms. Utters extremely loud squawk, 'rraaak-rraaak'. **SITES** Punta Leona Resort, and can be seen from boats in the evening on Rio Tárcoles.

Orange-fronted Parakeet ■ *Eupsittula canicularis* 23cm

DESCRIPTION Overall an all-green parakeet with long tail and orange patch on forehead and blue in middle of crown. Underparts paler green than upperparts, with olive tinge to breast; tail tipped with blue and bluish flight feathers. Large, whitish-yellow eye-ring, yellow eyes, whitish bill and grey feet. **HABITS AND HABITAT** Fairly common resident of northern Pacific lowlands to 1,000m. Favours open and semi-open habitat such as savannah, agricultural areas, secondary growth and forest edges. Often seen in noisy flocks of 30 or more birds, flying swift and direct. Feeds on fruits, seeds and flowers. Sounds a little like Finsch's Parakeet (see opposite), 'krey-krey-krey-krey'. **SITES** Guanacaste area, La Ensenada, Palo Verde and Santa Rosa.

Finsch's Parakeet

■ *Psittacara finschi* 28cm

DESCRIPTION Fairly large, mostly green, long-tailed parakeet that has bright crimson on forehead, patches of crimson on shoulder, crimson spots on sides of face and crimson thighs. Belly paler green-yellow and golden-yellow underwings in flight show red patches near shoulder. Orange eye surrounded by white eye-ring, and fairly large, whitish bill. Grey feet. **HABITS AND HABITAT** Common resident of Central Valley, Caribbean Slope and South Pacific Slope to 1,600m. Favours open habitat with scattered trees, farmland, parks and gardens. Usually seen in large, noisy flocks that roost in palms. Call a noisy screeching, 'key-key-key-key-key-key'. **SITES** Can be easily seen in hotel gardens and flying over San José.

Orange-chinned Parakeet ■ *Brotogeris jugularis* 18cm

DESCRIPTION Small, mostly green parakeet but with short, pointed tail. Small patch of orange on chin can be hard to see, so most easily identified by obvious brown patch on shoulder. Slightly bluish on head and rump, as well as on flight feathers. White eye-ring, brown eyes, pale pinkish bill and pinkish feet. **HABITS AND HABITAT** Common and widespread resident of lowlands on both Caribbean and Pacific Slopes to 1,200m. Favours open and semi-open areas with scattered trees, forest edges and secondary growth. Flies fast in noisy flocks of up to 50 birds, and feeds on fruiting trees, seeds, flowers and nectar. Call consists of various chirps and chattering 'zip-zip-zip-zip'. **SITES** Many suitable areas, including Palo Verde, Guanacaste.

Blue-headed Parrot

Pionus menstruus 24cm

DESCRIPTION Medium-sized parrot predominately green with deep blue head and upper breast. Black patch on cheek, reddish patch on centre of breast, bright red undertail and green belly. In flight shows bluish primary feathers. Pale eye-ring, dark eyes, black bill with pink patch and brownish feet. **HABITS AND HABITAT** Fairly common but localized resident of lowlands in southern Caribbean and also Pacific lowlands of San Vito and Golfito to 1,200m. Favours forest edges, open areas with scattered trees and wooded gardens. Eats fruits and seeds, and often seen in small flocks as well as larger communal roosts. Call a high-pitched shriek, 'squeek-squeek – squeek-squeek'. **SITES** Good spot to see it is around Las Esquinas Rainforest Lodge.

Red-lored Amazon

Amazona autumnalis 34cm

DESCRIPTION Fairly large and stocky, mostly green parrot. Distinguished by bright red patch on forehead and large amount of red on wings, which is especially easy to see in flight. Crown and hind-neck have tinges of blue, and there is some black scaling on neck and chest. Obvious creamy-white eye-ring, orange eyes, and two-tone yellow and blackish bill. Grey feet. **HABITS AND HABITAT** Common resident of wet lowlands on both Caribbean and Pacific Slopes to 1,000m. Favours forest edges, scattered trees and gardens, where it is usually seen in pairs feeding on fruits including figs, nuts and palms. Variety of calls, including loud, raucous 'klar-rik – klar-rik – ka-ka-ka'. **SITES** La Selva OTS.

Yellow-naped Amazon

■ *Amazona auropalliata* 36cm

DESCRIPTION Large, very stocky green parrot with big head and obvious bright yellow collar on back of nape. Slight tinge of blue on crown; belly paler green than upperparts. Red secondary flight feathers show best in flight; dark bluish-black primary tips and green tail appear more yellow-green on tail-band. Orange eyes surrounded by pale greyish eye-ring, and grey-black bill. Grey feet. **HABITS AND HABITAT** Uncommon resident in lowlands of North Pacific Slope down to Tárcoles and up to 600m. Favours deciduous forests and forest edges, and tall trees in savannah and farmland. Feeds on fruits including figs, seeds and flowering buds. Flies around in pairs and has a variety of calls, 'kraua-kraua'. **SITES** Palo Verde, Guanacaste.

Northern Mealy Amazon ■ *Amazona guatemalae* 38cm

DESCRIPTION Largest of the Costa Rican parrots and best distinguished by its plainness and obvious huge white eye-ring. Closer inspection shows a bluish tinge to top of head, red patches in secondary wing feathers in flights and dark bluish-black tips to wings. Hind-neck has a little blackish scaling, while outer half of tail is pale yellow-green. Orange eyes, horn-grey bill and grey feet. **HABITS AND HABITAT** Common resident of lowlands of Caribbean Slope and South Pacific Slope to 600m. Favours lowland wet forests, forest edges, and semi-open and secondary growth. Feeds on fruits including figs, nuts, palms and seeds, and often seen in flocks of 10 or more. Utters very noisy, squabbling squawks. **SITES** Carara NP and La Selva OTS.

Wedge-billed Woodcreeper

■ *Glyphorynchus spirurus* 15cm

DESCRIPTION Smallest of the woodcreepers; evenly brown upperparts, and slightly richer brown on wings, rump and uppertail. Buffy-brown underparts lightly streaked with cream, and pale buffy line above and behind eye. Indistinct mottling on face, and rufous tail with pointed tips. Short bill with sharp point, dark above and grey below. Greyish feet. **HABITS AND HABITAT** Common resident of lowlands of both Caribbean and South Pacific Slopes to 1,500m. Favours humid rainforests and older secondary growth, where it can be seen climbing vertically up tree trunks in search of tiny insects and spiders. Accompanies mixed species flocks. Song a descending series of notes, 'witt-wit-wit-wit-wt-wt-wt-wt-wt'. **SITES** Carara NP, La Selva OTS, Braulio Carrillo and La Tirimbina Lodge.

Northern Barred Woodcreeper ■ *Dendrocolaptes sanctithomae* 28cm

DESCRIPTION Large, heavily built woodcreeper with long, straight, stout dark bill. Overall plumage dark buffy-brown with more rufous on wings, rump and tail. Entire underparts, back and head covered in fine, narrow, uniform dark barring. This can be difficult to discern unless seen well. Dark eyes and grey legs. **HABITS AND HABITAT** Fairly common resident of lowlands and foothills of both Caribbean and Pacific Slopes to 1,200m. Favours mature forests, forest edges, clearings and semi-open areas with scattered trees. Feeds on insects including beetles and crickets, and small lizards, and also follows army ant swarms. Calls vary from long 'toowee-towee-toowee-toowee' to shorter 'tewy-tewy-tewy-tewy'. **SITES** La Selva OTS, Carara NP and Heliconias Lodge.

Cocoa Woodcreeper ■ *Xiphorhynchus susurrans* 22cm

DESCRIPTION Fairly large, brown-looking woodcreeper. Back and shoulders olive-brown, rump, tail and wings chestnut-brown, and underparts olive-brown with buff streaks from upper breast down to middle of belly. Neck, mantle and crown brown with buff spotting, buffy throat, and pale buff line over and behind eye. Bill fairly heavy and straight, with dark upper mandible and pale lower mandible. Greyish feet. **HABITS AND HABITAT** Common resident of both Caribbean and Pacific lowlands to 900m. Favours wet deciduous forests, forest edges, mangroves and clearings with scattered trees. Climbs tree trunks starting from near the ground. Feeds on insects including beetles and crickets, grubs and small lizards. Call a loud, rapid 'weet-weet-weet-weet-weet-weet-wit'. **SITES** Carara NP, La Selva OTS, Bosque del Rio Tigre and La Tirimbina Lodge.

Spot-crowned Woodcreeper

■ *Lepidocolaptes affinis* 20cm

DESCRIPTION Looks very similar to Cocoa Woodcreeper (see above), with main differences being a spotted not streaked crown (can be difficult to see), paler bill, and more heavily streaked underparts and back. Olive-grey feet. **HABITS AND HABITAT** Uncommon resident of highlands above 1,000m, where it replaces **Streak-headed Woodcreeper** *L. souleyetii*. Favours montane wet forests covered in mosses and epiphytes, where it forages like many woodcreepers by climbing up trunks and crawling along branches. Uses slender bill to extract grubs, beetles and crickets from behind bark and among lichen and mosses. Joins mixed species feeding flocks. Call a high-pitched 'deeeah', followed by descending trill, 'de-de-de-de-de-de-de-de-de'. **SITES** Savegre Mountain Lodge and Paraiso Quetzales.

Fasciated Antshrike ■ *Cymbilaimus lineatus* 18cm

DESCRIPTION Similar to Barred Antshrike (see opposite), but male is completely covered with fine white barring over black body. Crown black extending to nape. Bright red eyes, and hooked bill has dark upper mandible and pale grey lower mandible. Pale grey legs. Female black-brown and entirely finely barred with buff, with rich rufous crown. **HABITS AND HABITAT** Fairly common resident in Caribbean lowlands and foothills to 1,200m. Favours thick tangles and vines in forest edges and secondary growth. Elusive and hard to see, often sitting quietly for long periods. Feeds mostly on insects. Call comprises 4–6 notes, 'whoo-whoo-whoo-whoo'. **SITES** La Selva OTS is a reliable spot in which to see it.

Great Antshrike ■ *Taraba major* 20cm

DESCRIPTION Male has all-black upperparts, including head and tail; wings have three narrow white wing-bars, and underparts are all white from throat to lower belly. Undertail and flanks are sooty-grey. Heavy black bill with hooked tip, bright red eyes and bluish-grey legs. Female similar to male except that black of male is replaced with rich rufous-chestnut, and flanks have rufous wash. **HABITS AND HABITAT** Fairly common resident of lowlands on Caribbean Slope and South Pacific Slope to 1,100m. Favours forest edges, secondary growth, stream edges and thick cover, especially vine tangles where it is hard to see. Feeds on variety of insects and small lizards. Call a rapidly descending 'top – top – top – top-top-top-top-top-tp-tp-tp-tp–ggrrrrrrrrrrr'. **SITES** La Selva OTS and Arenal area.

Barred Antshrike ■ *Thamnophilus doliatus* 16cm

DESCRIPTION Similar to Fasciated Antshrike (see opposite), but male has much broader white barring over black body. Male also has shaggy black-and-white striped crest and pale yellow eyes. Female very different from similar Fasciated Antshrike in being plain rufous brown above and cinnamon below, with only a small amount of streaking on sides of face. Also has a rufous crest. Greyish bill and legs. **HABITS AND HABITAT** Common resident of both Pacific and Caribbean lowlands and Central Valley to 1,400m. Favours forest edges, thickets and understorey, where it can remain elusive. Often in pairs. Feeds on variety of insects. Call a 'chaaarrg'; song 'grr-grr-grr-grr-gr-gr-gr-gr-gr-g-gwiikk'. **SITES** Many sites, including Las Esquinas, La Ensenada and Carara NP.

Black-hooded Antshrike ■ *Thamnophilus bridgesi* 15cm

DESCRIPTION Male has dark sooty-black upperparts, head, tail and most of underparts. Lower belly and undertail dark grey, and small, clean white spots on shoulder and wing. Female browner that male, with blackish head and tail, and heavily streaked white head and underparts. Dark eyes, black bill with hooked tip, and grey legs. **HABITS AND HABITAT** Common resident of southern Pacific lowlands to 1,200m. Favours forest edges and thickets in secondary growth, where it keeps low and follows mixed species flocks. Feeds on variety of insects and spiders. Call similar to Great Antshrike's (see opposite), 'dpp-dpp-dpp-dp-dp-dp-dp-dp-dp–ggrrrrrrrrrrr'. **SITES** Easily spotted in Carara NP.

Plain Antvireo

■ *Dysithamnus mentalis* 11cm

DESCRIPTION Rather stout and rounded, with shortish tail and quite heavy, slightly hooked bill. Male has greyish head and back, olive wings with thin white wing bars, olive tail, pale whitish throat, grey wash on breast and pale yellow underparts. Dark grey face encircled by pale grey line. Female's upperparts olive-brown; underparts pale yellowish with brownish wash on breast; white throat, grey face and chestnut crown. Grey bill and legs. **HABITS AND HABITAT** Fairly common resident of Caribbean and southern Pacific foothills to 2,000m. Prefers to stay low in understorey of mature forests, where it joins mixed flocks and feeds on insects. Song an accelerating 'ha - ha - ha-ha-ha-ha-ha-ha-ha-haa'. **SITES** Carara NP, Las Esquinas and Bosque del Rio Tigre.

Dot-winged Antwren ■ *Microrhopias quixensis* 10cm

DESCRIPTION Small and slender; male jet black with broad white wing-bar and white spots on shoulder. Graduated tail has broad white tips to each feather. Female has same wing pattern and tail as male, but underparts chestnut, and head and back slate-grey. Black eyes, thin dark bill and dark legs. **HABITS AND HABITAT** Common resident of southern Pacific lowlands, and uncommon in Caribbean lowlands to 1,000m. Favours understorey and vine tangles of mature wet forests and forest edges, and dense secondary growth, where it generally travels around in pairs. Joins up with mixed species flocks and feeds on small insects gleaned from foliage. Song a series of high-pitched, ascending whistles that drop off on last note, 'cip-cip-cip-cip-cip-cip-cip-ceeer'. **SITES** Best place to see it is Carara NP.

Male

Female

Chestnut-backed Antbird ■ *Myrmeciza exsul* 14cm

DESCRIPTION Heavy bodied and dumpy. Head and underparts slaty-black; paler on lower belly and undertail. Upperparts, wings and tail chestnut; large obvious area of pale blue orbital skin. Female and male of Pacific race mostly rufous all over, but still have pale blue orbital skin. Blackish-grey bill and legs. **HABITS AND HABITAT** Common to fairly common resident of Caribbean and southern Pacific lowlands and foothills to 1,200m. Favours mature wet forests, where it keeps low to the ground among thickets, and forages on the ground among leaf litter for insects and grubs. Pairs sometimes follows army ants and mixed flocks. Regular call 2–3 hollow notes, 'peeet-peeew'. **SITES** Carara NP, La Selva OTS and Las Esquinas Rainforest Lodge.

Spotted Antbird ■ *Hylophylax naevioides* 11cm

DESCRIPTION Small and plump with short tail. Male has slaty-grey head with black throat, rich chestnut back, black wings with two broad chestnut wing-bars, and clean white underparts with large black spots on breast. Female a duller buffy-brown version of male. Blackish bill and grey legs. **HABITS AND HABITAT** Fairly common to uncommon in lowlands of Caribbean Slope to 1,000m. Favours understorey of mature and secondary wet forests, where it forages on or near the ground and readily perches on vertical twigs. Often follows army ant swarms. Call 'psip-psip'; song a descending series of buzzy notes, 'peter-peter-peter-peter-peter-peter-peter-peet'. **SITES** Arenal Observatory Lodge and Heliconias Lodge.

Greenish Elaenia ■ *Myiopagis viridicata* 13cm

DESCRIPTION Plain and nondescript flycatcher with no obvious field marks. Uniformly olive-green upperparts and pale yellow underparts with very faint streaking. Greyish throat and upper breast, plain wings and dusky-green with pale edges to secondary feathers only. Indistinct pale greyish stripe over eye, black bill, dark eyes and dark greyish legs. **HABITS AND HABITAT** Fairly common resident throughout lowlands of Pacific Slope to 1,500m. Favours forest edges, secondary growth, and open areas with scattered trees and bushes. Usually seen alone foraging high in canopy for insects. Also feeds on berries. Call a sharp 'pseeew' or 'pseewit'. **SITES** Carara NP is a good spot in which to see it.

Yellow-bellied Elaenia ■ *Elaenia flavogaster* 15cm

DESCRIPTION Upperparts olive-green, throat and breast pale grey, belly and undertail pale yellow. Two pale buffy-white wing-bars, pale edges to flight feathers, narrow white eye-ring, and shaggy long crest that when raised shows central white crown-patch. Dark bill with pinkish base of lower mandible. Dark legs. **HABITS AND HABITAT** Common and widespread resident in lowlands and foothills over much of Costa Rica to 2,000m. Favours open and semi-open habitat, secondary growth and gardens. Catches insects on short flights, and also eats berries and other small fruits. Very active and often noisy species regularly seen with crest raised. Call a wheezy 'weeeerr'; song a wheezy 'cher-beer'. **SITES** May be seen anywhere within lowlands.

Mountain Elaenia ■ *Elaenia frantzii* 15cm

DESCRIPTION Upperparts dull olive-green; wings duller olive-grey with two distinct whitish wing-bars formed by large spots. Flight feathers edged yellowish, underparts pale yellow-grey and tail dusky-olive. Head looks very rounded and has no crest; thin white eye-ring, and pale line between eye and bill (lores). Dark bill with pink-grey base of lower mandible. **HABITS AND HABITAT** Common resident of highlands, where it is found above 1,200m to timberline. Favours forest edges, secondary growth, and scattered trees and gardens. Catches insects on short flights, but mostly feeds on berries and seeds. Call a sharp 'pseeew'. **SITES** Savegre Mountain Lodge, Paraiso Quetzales and Tapantí NP.

Southern Beardless Tyrannulet ■ *Camptostoma obsoletum* 10cm

DESCRIPTION Very small, greyish olive-green bird with distinct shaggy crest. Dusky grey wings with two pale yellow-white wing-bars and pale whitish edges to flight feathers. Tail also dusky-olive, with whitish tip. Thin white eye-ring and pale area forming spectacle around eye. Throat pale greyish, breast washed with olive and pale yellowish underparts. Tiny black bill and dark legs. **HABITS AND HABITAT** Fairly common resident in lowlands of South Pacific Slope to 1,000m. Favours open forests, clearings, secondary growth and gardens, where it can often be seen hovering or picking small insects from leaves. Often cocks tail. Song comprises 4–6 descending notes, 'sweee-swee-swee-swee-swe-swe-swe-swe-swe'. **SITES** Carara NP and Bosque del Rio Tigre.

Torrent Tyrannulet ■ *Serpophaga cinerea* 10.5cm

DESCRIPTION Tiny grey-and-black flycatcher entirely restricted to mountain streams. Hind-neck, back and rump pale grey, underparts very pale grey-white, tail and wings blackish with distinct white wing-bars, and crown and cheeks sooty black. Tiny black bill and black legs. **HABITS AND HABITAT** Common resident of highland mountain streams at 500–2,200m. Nearly always near swift rocky streams, most often seen perched on rocks, where it can be seen pumping its tail up and down. If not perched on rocks in a stream, it will be close to the bank or maybe on small overhanging branches. Feeds on damselflies and other small insects. Call a sharp, shrill rapid 'ptsr – psit-psit-psit-psit-psit-psit-psit-pt-pt-pt-pt'. **SITES** Streams by Trogon Lodge, Savegre Mountain Lodge and Bosque de Paz.

Yellow Tyrannulet ■ *Capsiempis flaveola* 10.5cm

DESCRIPTION Small and slender with longish tail, recalling a warbler. Upperparts yellow-olive with two distinct yellow wing-bars; underparts bright yellow. Bright yellow stripe over eye and dark line on either side of eye. Small, slender black bill and pale greyish-pink legs. **HABITS AND HABITAT** Fairly common resident of lowlands of Caribbean and South Pacific Slopes to 1,200m. Favours low thickets and shrubs, bushes and small trees, and coffee plantations, where it often stays well hidden. Generally in pairs and feeds on small insects and beetles, occasionally flycatching. Call a bubbling 'bibipoe-bibipoe-bibipoe'. **SITES** La Selva OTS area and Las Esquinas Rainforest Lodge.

Mistletoe Tyrannulet ■ *Zimmerius parvus* 9.5cm

DESCRIPTION Very small, olive-grey flycatcher that is most identifiable by very obvious yellow edges to wing feathers (no wing-bars). Rest of upperparts grey-green, tail dusky. Underparts with throat and breast greyish-white with fine grey streaks; white belly and flanks with yellowish wash. Light grey-white stripe over eye, pale yellow eyes, and tiny black bill with brown at base of lower mandible. Grey legs. **HABITS AND HABITAT** Common and widespread resident throughout much of Costa Rica to 3,000m. Found in most habitats, including gardens and secondary growth, forest edges, pastures with scattered trees and bushes, where it feeds mostly on mistletoe berries as well as small insects. Call a loud 'peeeew'; song 'derr-d-dee – derr-d-dee'. **SITES** La Selva OTS and Rancho Naturalista.

Scale-crested Pygmy Tyrant ■ *Lophotriccus pileatus* 8cm

DESCRIPTION Tiny, plump flycatcher. Upperparts olive-green, including wings and tail. Wings have two buffy-yellow wing-bars. Underparts yellowish-olive on belly and breast; pale throat. Both throat and breast finely streaked with grey, and tawny-olive face. Crown crest appears mostly pink with black tips – this is obvious even when crest is lying flat. Pinkish-white eyes, bill has black upper mandible and pink lower mandible, and legs pale greyish. **HABITS AND HABITAT** Fairly common resident of foothills and lower mountain slopes of both Caribbean and southern Pacific areas at 300–1,700m. Prefers to be in middle elevation of wet, mossy forests, forest edges and secondary growth. Feeds mostly on insects. Call a rising, high-pitched 'trrrip – tip-tip-tip-tip-tp-tp-ttttttttttttttttt-trip'. **SITES** Las Esquinas Rainforest Lodge and Arenal Observatory Lodge.

Common Tody-flycatcher ■ *Todirostrum cinereum* 10cm

DESCRIPTION Very small with large head. Bright yellow underparts from throat right down to undertail. Crown, nape and back sooty-grey, and forecrown and face black. Rest of upperparts olive-green with bright yellow edges to all wing feathers; tail graduated blackish and with white edges. Bright yellow eyes, rather long, flattened bill black above and paler below, and blue-grey legs. **HABITS AND HABITAT** Common and widespread resident over much of Costa Rica's lowlands to 1,500m. Prefers to keep low in forest edges, gardens, coffee plantations, secondary scrub and semi-open areas. Active little bird that constantly wags its tail. Feeds on insects. Call a rapid 'prriirit-prriirit'. **SITES** Many places, including La Selva OTS, Carara, Cerro Lodge and Bosque del Rio Tigre.

Yellow-olive Flatbill ■ *Tolmomyias sulphurescens* 13cm

DESCRIPTION Fairly small and overall looks olive-green with grey head. Most important feature in identification is pale whitish eyes. Upperparts olive-green with bright yellow edges to wing feathers, yellow belly and undertail, and greyish head, throat and upper breast. Pale grey line between bill and eye, whitish eye-ring and small, wide, flat greyish bill. Grey legs. **HABITS AND HABITAT** Common and widespread resident of both Caribbean and Pacific lowlands to 1,400m. Favours dry forest edges, secondary growth, open woodland and gardens, where it feeds on insects caught in the air or gleaned from foliage, and berries. Call comprises short, sharp, sparrow-like 'tsip' and sharp, buzzy 'bzeeeek'. **SITES** La Selva OTS, Carara NP and La Ensenada.

Black Phoebe ■ *Sayornis nigricans* 17cm

DESCRIPTION Medium-sized flycatcher that appears to be mostly black. Upperparts, head, breast, flanks and tail matt black. Belly clean white and undertail sooty-grey. Longish tail has white edges, and black wings have pale whitish edges to flight feathers and two greyish-white wing-bars. Rather square headed with black bill and legs. **HABITS AND HABITAT** Common resident in Central Valley and South Pacific Slope at 500–2,200m. Almost always near water, especially fast-running, rocky streams, where it perches on rocks, wires and fences, before flycatching for a variety of insects including dragonflies. Often roosts on buildings. Call a sharp 'pseee'; song 'pe-bee – pe-bee'. **SITES** Savegre Mountain Lodge, Trogon Lodge and Bosque de Paz.

Northern Tufted Flycatcher ■ *Mitrephanes phaeocercus* 12cm

DESCRIPTION Small flycatcher that looks overall cinnamon-buff and has obvious pointed crest. Upperparts warm olive-brown, including crest and tail. Dusky wings with two pale buffy wing-bars and white edges to tertials. Breast to lower belly rich cinnamon, turning to buffy on undertail. Pale ochre throat and spot in front of eye. Thin yellowish eye-ring. Tiny bill has black upper and yellow lower mandible, and legs are dark. **HABITS AND HABITAT** Common resident in highlands on both Caribbean and Pacific Slopes at 1,200–3,000m, occasionally lower. Favours montane wet forests, forest edges, secondary growth and clearings. Typically flycatches from exposed dead twig, often returning to the same perch. Most often heard call is 'wit-wit-wit-wit-wit-wit-wit-wit'. **SITES** Savegre Mountain Lodge and Trogon Lodge.

Dark Pewee ■ *Contopus lugubris* 17cm

DESCRIPTION Large, mostly dark sooty-grey flycatcher. Overall sooty-grey darkest on head, upperparts, wings and tail. Breast sooty-grey, becoming paler on lower belly, and throat pale grey. Distinctive crest forming point at back of head, no wing-bars, dark upper mandible and orange lower mandible, and dark legs. **HABITS AND HABITAT** Fairly common resident although sparse in highlands at 1,200–2,250m. Favours wet montane forest edges near streams with exposed tall trees, in which it perches high and from which it flycatches, returning to same perch. Call a repeated 'quip-quip – quip-quip'. **SITES** Best in valley around Savegre Mountain Lodge and Trogon Lodge.

Tropical Pewee ■ *Contopus cinereus* 13cm

DESCRIPTION Small, dusky-olive-brown flycatcher that looks similar to Western and Eastern Wood Pewees (*C. sordidulus* and *C. virens*). Best distinguishing features include darker cap, yellowish wash to belly, pale spot in front of eye (lores), and yellow lower mandible with black tip. Otherwise pretty much olive-brown all over with distinct whitish edges to wing feathers and two pale greyish wing-bars. Black upper mandible and dark legs. **HABITS AND HABITAT** Common resident of Caribbean lowlands to 900m, and uncommon in Central Valley and southern Pacific lowlands to 1,200m. Prefers to perch low in forest edges, hedgerows, gardens and parks. Often perches on wires. Catches insects in flight. Call a loud 'swiiip-swiiip-swiiip'. **SITES** La Selva OTS, Braulio Carrillo and La Quinta Country Inn.

Yellowish Flycatcher

■ *Empidonax flavescens* 12.5cm

DESCRIPTION As its name suggests, a very yellowish-looking flycatcher. Upperparts yellow-olive, more dusky on wings and tail. Two buffy-yellow wing-bars, yellow throat and belly, and breast tinged with orange. Slight crest on head; good identification feature is triangular yellow spot behind eye. Thin yellow eye-ring, bill black above and orangey below, and grey legs. **HABITS AND HABITAT** Common resident in highlands throughout Costa Rica at 800–2,500m. Prefers to perch very low and can be found in montane forest edges and secondary growth, even gardens, where it flycatches for small insects. Call a sharp 'sweeeew'. **SITES** Savegre Mountain Lodge, Trogon Lodge and Paraiso Quetzales.

Black-capped Flycatcher ■ *Empidonax atriceps* 11cm

DESCRIPTION Very small, sooty-grey flycatcher with dark blackish head and distinct whitish eye-ring extending slightly behind eye. Upperparts sooty-grey, including tail and wings, which have two narrow buffy wing-bars; underparts pale buffy-grey with darker wash on breast and pale throat. Upper mandible black and lower mandible yellow-orange with dark tip. Blackish legs. **HABITS AND HABITAT** Common resident of highlands and mountains mostly above 2,000m to treeline. Prefers to stay low in montane forest edges, bushes, scrubby pastures, gardens and paramo. Flycatches from regular perch and quivers tail on landing. Call a sharp, repetitive 'pwit'. **SITES** Savegre Mountain Lodge, Trogon Lodge and Paraiso Quetzales.

Long-tailed Tyrant

■ *Colonia colonus* 14cm

DESCRIPTION Small, sooty-black flycatcher with broad white stripe over and around eye, and grey-white back and rump. Two very long central tail feathers add another 10cm to size. Wings, tail and underparts all sooty-black, and crown grey. Tiny black bill and blackish legs. Female slightly greyer all over than male, with a fractionally smaller tail. **HABITS AND HABITAT** Fairly common resident of Caribbean lowlands to 600m. Favours forest edges, clearings, and open and semi-open areas, where it sits on highest dead vertical twig or branch generally above the surrounding canopy. Flycatches and returns to regular perch. Call a whistled, ascending 'sweeeep'. **SITES** Arenal Dam area, La Selva OTS and Caño Negro area.

Piratic Flycatcher ■ *Legatus leucophaius* 15cm

DESCRIPTION Medium-sized, streak-breasted, brownish flycatcher. Upperparts dull olive-brown, and wings dusky with pale buffy-white wing-bar. Underparts pale creamy-white with brownish streaking on breast to mid-belly. Head pattern with brown face, white stripe over eye to back of head, broad white stripe on cheek, thin brown malar stripe and whitish throat. Small black bill and dark legs. **HABITS AND HABITAT** Common breeding migrant in February–September in lowlands of both Caribbean and Pacific Slopes to 1,600m. Favours open areas and forest edges, typically perching high on tree tops. Parasitic, taking over other species' nests. Most regular call from tree top is 'sweeyee'. **SITES** La Selva OTS, Carara NP and Bosque del Rio Tigre.

Social Flycatcher ■ *Myiozetetes similis* 16cm

DESCRIPTION One of several similar-looking species, with small bill and greyish (not black) face. Upperparts including tail and wings dull olive-green; wing feathers with narrow yellow edges. Bright yellow underparts, white throat, broad white eye-stripe that does not meet on nape, and mostly concealed red-orange crown-patch. Small black bill and black legs. **HABITS AND HABITAT** Common and widespread resident over much of Costa Rica to 2,000m. Found in wide variety of habitats, including open pastures with scattered trees, secondary growth, river edges and gardens. Regularly flycatches for insects even low over water, and also feeds on fruits including berries. Call like squeaky toy, 'pseeear'. **SITES** Almost anywhere, including Rancho Naturalista, La Ensenada and La Selva OTS.

Great Kiskadee

■ *Pitangus sulphuratus* 23cm

DESCRIPTION Large flycatcher with black-and-white striped head pattern. Upperparts olive-brown, wings and tail with rufous edges, flight feathers especially rufous on primaries. Bright yellow underparts; white throat contrasts with black face-mask and broad white stripe that runs above eye and meets at back of nape. Sides of crown black with bright orange-yellow in centre of crown. Fairly long, stout black bill and black legs. **HABITS AND HABITAT** Common and widespread resident throughout lowlands of much of Costa Rica to 1,500m. Favours open habitat with scattered trees, secondary growth, parks and gardens, often near water. Very noisy and restless. Feeds on insects, lizards, frogs and worms. Call a loud and distinctive 'kis-ka-dee'. **SITES** Many places, including La Selva OTS, Palo Verde, La Quinta and Cerro Lodge.

Streaked Flycatcher

■ *Myiodynastes maculatus* 20cm

DESCRIPTION Upperparts brown heavily streaked with black, and cinnamon rump; wings dusky with primary feathers rufous, and tail also rufous. Underparts whitish with heavy streaks from throat to lower belly. Head pattern shows brown. Streaked head with broad white stripe above eye, and broad black mask through eye with another broad white stripe below this. Stout bill black on upper mandible and pale pinkish on lower mandible, with black tip. Dark grey legs. **HABITS AND HABITAT** Fairly common resident of lowlands on Pacific Slope to 1,400m. Favours forest edges, semi-open areas, mangroves and gardens. Catches insects in flight. Call a sharp 'chic'. **SITES** La Selva OTS and Carara NP.

Boat-billed Flycatcher ■ *Megarynchus pitangua* 23cm

DESCRIPTION Large flycatcher with black-and-white head pattern. Very similar in looks to Great Kiskadee (see p. 107), with most notable differences being an even larger, wider black bill and no rufous in wings or tail, just dull olive. Rest of upperparts olive-brown, white throat, bright yellow underparts, black-and-white striped head, and white stripe over eye not quite meeting on nape. Yellow-orange crown-stripe and dark legs. **HABITS AND HABITAT** Common and widespread resident over much of Costa Rica to 2,200m. Favours forest edges, dry savannah, secondary growth and semi-open areas. Feeds on variety of insects including cicadas; also fruits including berries. Chases other birds. Call a nasal 'prrrrrrrrrraaah'. **SITES** Carara NP, Hacienda Solimar and Palo Verde.

Tropical Kingbird

■ *Tyrannus melancholicus* 20cm

DESCRIPTION Upperparts greyish-olive, dusky wings tinged with pale edges and no wing-bars; yellow underparts, pale grey-white throat and pale grey head with concealed orange crown-patch. Dark grey band through eye, and black bill and legs. **HABITS AND HABITAT** Common resident – in fact one of the most common birds in Costa Rica, found throughout the country to 2,000m. Favours open and semi-open areas including savannah, farmland, river edges, towns, parks and gardens, as long as it can find exposed perches. Often seen on tops of dead trees and utility wires. Flycatches for insects, including butterflies and dragonflies, and will eat berries. Regularly chases raptors. Call a sharp, bubbling 'prt-prt-prt-prprprprp-prrrrt-prrrrt-prt-pt'. **SITES** Can be found just about anywhere.

Scissor-tailed Flycatcher

■ *Tyrannus forficatus* 19cm, plus up to 15cm tail

DESCRIPTION Very long-tailed, greyish flycatcher. Pale grey back, wings black with grey edges to all feathers, very elongated, forked tail black with whitish edges, pink rump, pale grey head and underparts, and flanks, undertail and underwing salmon-pink. Dark line through eye, and black bill and dark legs. **HABITS AND HABITAT** Common winter resident in October–April, mostly along Pacific Slope to 2,200m. Favours dry savannah, open grasslands and fields, where it perches on scattered trees, bushes and fence wires. Hundreds often seen in evenings going to roost in marshes. Feeds on flying insects and berries. Call a squeaky twitter, like a dog chewing squeaky toy. **SITES** Guanacaste, Palo Verde, Santa Rosa and Hacienda Solimar.

Rufous Mourner

Rhytipterna holerythra 20cm

DESCRIPTION Entirely rufous with richer chestnut-cinnamon on head and upperparts, including wings and tail. Slightly paler cinnamon-buff on underparts with darker wash on breast. Black-brown eyes, and bill has dark upper mandible with pink at base and lower mandible mostly pinkish with dark tip. Grey legs. **HABITS AND HABITAT** Common resident in lowlands and foothills of both Caribbean and Pacific Slopes to 1,200m. Prefers to stay high up in mature rainforests, as well as tall secondary growth. Can sit motionless for ages and feeds on variety of insects and caterpillars. Call a loud, drawn-out whistle, 'swweeet-heeaarrr'. **SITES** Bosque del Rio Tigre, Rancho Naturalista and Heliconias Lodge.

Dusky-capped Flycatcher

Myiarchus tuberculifer 16cm

DESCRIPTION Fairly small flycatcher with best distinguishing feature being its contrasting blackish-grey head. Olive-brown upperparts; dusky wings and tail with warm brown-rufous edges to flight feathers producing indistinct wing-bars. Greyish throat to upper breast, yellow belly and undertail, dark eyes, and black bill and legs. **HABITS AND HABITAT** Common resident over much of Costa Rica to 1,800m. Favours open edges of forests, secondary growth, scattered trees, mangroves and gardens. Tends to perch on low, exposed branches and tangles, and feeds by flycatching insects, but will also eat grasshoppers and berries. Call a simple whistle, 'weeeee'. **SITES** Arenal Observatory Lodge, La Selva OTS and La Ensenada.

Great Crested Flycatcher ■ *Myiarchus crinitus* 20cm

DESCRIPTION Rather large flycatcher mostly distinguished from similar flycatchers by very bright yellow underparts contrasting sharply with darkish grey upper breast and throat. Also pale pinkish base to lower mandible. Rest of bill dark. Green-olive back, dusky wings with pale wing-bars, bright rufous primaries, tail mostly rufous, brownish-olive head with shaggy crest and grey on face. Dark legs. **HABITS AND HABITAT** Common passage migrant and winter resident on both Caribbean and Pacific Slopes in September–April to 1,800m. Found in many habitats, such as woodland, open areas with scattered trees, mangroves and gardens. Catches insects in flight and also feeds on berries. Call a loud, rising 'wheeeeet'. **SITES** Many places, including La Ensenada and La Selva OTS.

Brown-crested Flycatcher ■ *Myiarchus tyrannulus* 19.5cm

DESCRIPTION Very similar to Great Crested Flycatcher (see above), but less olive and more brownish above, with short crest, and pale yellow underparts not bright below. Also all-dark bill. Pale grey on throat and breast more extensive; otherwise all other features are similar – that is, rufous in wings and tail, pale wing-bars and dark legs. **HABITS AND HABITAT** Common resident of northern Pacific lowlands and foothills to 900m. Prefers to keep low to mid-height in tangles, scattered bushes and trees, semi-open areas, parks and gardens. Catches insects in flight and often seen returning to open perch. Most often heard call a repeated sharp 'huit'. **SITES** Palo Verde, La Ensenada and Santa Rosa.

Northern Royal Flycatcher

Onychorhynchus mexicanus 17cm

DESCRIPTION Upperparts mostly dull brown with richer, warmer brown-rufous on rump and tail. Wings covered with smallish buff spots, throat pale whitish-buff and underparts warm buffy-yellow. Interesting elongated head with protruding flat crest. This is rarely raised but if seen is spectacular, being an orange-red fan, with black-and-blue tips. Pale line between bill and eye, pale eyes and fairly long, darkish bill. Orangey-brown legs. **HABITS AND HABITAT** Fairly common to uncommon resident in lowlands of both Caribbean and Pacific Slopes to 900m. Can be found in dense, shady forests, where it prefers to be near streams or dried-up river beds. Call a harsh, loud 'keeeeyup'. **SITES** Carara NP and Santa Rosa.

White-collared Manakin

Manacus candei 10.5cm

DESCRIPTION Tiny and plump with short tail. Male has black cap; white on rest of head, throat, upper breast and back forms broad white band. Bright yellow belly and undertail, olive-green rump and uppertail, and black wings. Tiny black bill and orangey legs. Female all green but still has orangey legs. **HABITS AND HABITAT** Common resident of Caribbean lowlands and foothills to 900m. Favours thickets in secondary growth, humid forests and forest edges, where it stays low and is quite secretive. Small groups gather in leks where they fly back and forth making snapping sounds with their wings. Feeds on fruits including berries. Call a high-pitched 'prrreeew'. **SITES** La Selva OTS, La Tirimbina and Rancho Naturalista.

Orange-collared Manakin ■ *Manacus aurantiacus* 10.5cm

DESCRIPTION Tiny and plump with short tail. Male looks pretty much the same as White-collared Manakin (see opposite), except that it has a broad orange band instead of a yellow one. Still retains black cap, black wings and small black patch on back, and bright olive-green rump and uppertail. Tiny black bill and orange legs. Female green with orange legs. **HABITS AND HABITAT** Fairly common resident of central and southern Pacific lowlands to 1,100m. Favours mature wet forests, forest edges, shady understorey and secondary growth. Stays low and gathers in small groups at leks, where males perform dances and jump back and forth on small branches, making snapping sounds with their wings. Feeds on insects and berries. Call a loud, rippling 'chrrrrrrrraa'. **SITES** Best spots to see it are Carara NP and Las Esquinas.

Red-capped Manakin ■ *Dixiphia mentalis* 10cm

DESCRIPTION Tiny and plump with short tail. Male mostly jet black with obvious bright red head. Pale yellow chin, bright yellow thighs, tiny, pale pinkish bill and pinkish legs. Female dull olive with greyish throat and paler belly, darker brown eye, dark bill and brownish legs. **HABITS AND HABITAT** Fairly common and local resident of both Caribbean and Pacific lowlands to 1,100m. Favours mature wet forests, forest edges, secondary growth and even gardens, staying at low to middle elevations and feeding on fruits including berries. Forms leks where males dance and slide on horizontal branches. Call a sharp 'tsip'; song a 'tsip-tsip-tsweeeeeeeeeee'. **SITES** La Selva OTS and Las Esquinas.

Masked Tityra

■ *Tityra semifasciata* 20cm

DESCRIPTION Male mostly white with greyish tinge. Outer half of wing black, and broad black band towards end of white tail. Red skin around eye and up to edge of bill, all surrounded by black face. Reddish-brown eyes and thickish red-pink bill with black tip. Blue-grey legs. Female similar to male but has less black on face and sooty-brown upperparts. **HABITS AND HABITAT** Common and widespread resident over much of Costa Rica to 2,000m. Favours forest edges, gardens, and open and semi-open areas with scattered trees. Regularly found in small groups high on fruiting trees. Feeds on fruits including berries, insects and small lizards. Call like high-pitched mechanical frog, 'g-reek – g-reek – g-reek'. **SITES** La Selva OTS and Carara NP.

Rufous-browed Peppershrike ■ *Cyclarhis gujanensis* 15cm

DESCRIPTION Fairly stout with big head and heavy bill. Upperparts olive-green including tail; underparts mostly yellow fading to white on undertail. Yellow throat, and rest of head pale grey with obvious rufous brow over eye. Pale orange eyes, silvery-blue bill with hooked tip, and pinkish-grey legs. **HABITS AND HABITAT** Fairly common resident of Central Highlands and uncommon in Pacific lowlands, mostly at 700–2,500m. Favours forest edges, semi-open areas, secondary growth, mangroves and gardens. Slow moving and difficult to see, especially among foliage in tree tops. Feeds on insects, spiders and small lizards. Song a loud 'do-yoo-wash-every-week'. **SITES** Trogon Lodge and Tárcoles mangroves.

Red-eyed Vireo ■ *Vireo olivaceus* 14cm

DESCRIPTION Upperparts olive-green including tail. Underparts dull white washed with pale yellow on flanks and undertail. Greyish-looking head with strong white eyebrow bordered by black. Bright red eyes and blue-grey bill and legs. Similar **Yellow-green Vireo** *V. flavoviridis* is brighter green with more obvious yellow on underparts. **HABITS AND HABITAT** Common passage migrant throughout lowlands of much of Costa Rica in August–November and March–May to 2,000m. Occurs pretty much anywhere with trees, often high up, and freely mixes with flocks of other species. Feeds on insects and berries. Generally silent but sometimes sings repeatedly, 'cheevi-cheerio'. **SITES** Found anywhere.

Lesser Greenlet ■ *Hylophilus decurtatus* 10cm

DESCRIPTION Small and rather short tailed. Upperparts olive-green, including wings and tail. Underparts pale whitish with yellow wash to flanks, breast and undertail. Throat white and head pale grey with obvious white eye-ring surrounding dark eye; bill can appear pale creamy-grey and legs grey. **HABITS AND HABITAT** Common and widespread resident throughout lowlands and foothills to 1,400m. Prefers to stay fairly high up in forest edges, secondary growth and semi-open areas, where it actively seeks insects and spiders among the foliage, often hanging upside down. Also eats berries. Song a 'che-checheree'. **SITES** La Selva OTS, La Tirimbina and Carara NP.

Brown Jay ■ *Psilorhinus morio* 38cm

DESCRIPTION Large, mostly brown jay. Upperparts dark brown, including wings; head and tail darker with tail having graduated white tip to each feather. Dark brown head continues to breast, where it fades to whitish belly and undertail. Bill black in adults, while juveniles have yellow bill and yellow eye-ring. Dark legs. **HABITS AND HABITAT** Common resident found over much of Costa Rica except southern Pacific area. Occurs in lowlands and middle elevations to 2,400m. Favours deforested areas, banana and coffee plantations, and open areas with scattered trees. Forages in groups of 6–10 for insects, small lizards and fruits at all levels from ground to tree top, and readily comes to birdfeeders. Loud call, 'piyaaah-piyaaah'. **SITES** Feeders at Arenal Observatory Lodge and Rancho Naturalista.

White-throated Magpie-jay ■ *Calocitta formosa* 46cm

DESCRIPTION Large blue-and-white jay with long crest and very long tail. Upperparts and rump grey-blue, wings and nape bright blue-grey, and tail bright blue with large white tips. Underparts white, including throat and face; black line circles rear of face, then continues to make narrow black breast-band. Long, forwards-curling black crest, and black bill and legs. **HABITS AND HABITAT** Common resident found mainly in lowlands of northern Pacific area to 1,200m. Favours forest edges and semi-open areas with scattered trees and bushes. Moves around in noisy flocks of 5–10, hunting for variety of insects, small lizards and fruits. Call mostly a harsh, loud 'kaaarr-kaaarr'. **SITES** Arenal area, La Ensenada and Palo Verde.

Long-tailed Silky-flycatcher ■ *Ptiliogonys caudatus* 24cm

DESCRIPTION Slender green-and-grey bird with pointed crest and long, pointed tail. Male has blue-grey back, rump, breast and forehead, with rest of plumage, including neck, throat, lower belly and undertail, bright yellow-olive. Long tail blackish underneath, with large white patch and two small, pointed central feathers. Bright yellow eye-ring, small black bill and black legs. **HABITS AND HABITAT** Fairly common resident of central southern highlands above 1,500m to timberline. Favours forest edges, tall trees and gardens, where pairs usually sit high on exposed tree tops. Flycatches for insects and eats berries. Call like a cricket's, 'prrrtt-prrrtt-prrrtt'. **SITES** Savegre Mountain Lodge, Trogon Lodge and Paraiso Quetzales.

Mangrove Swallow ■ *Tachycineta albilinea* 13cm

DESCRIPTION Similar to **Blue-and-white Swallow** *Notiochelidon cyanoleuca* but more metallic green-blue with obvious white rump. Top half of head, back, rump and shoulders metallic green-blue, which looks bluer over water. Flight feathers dusky-black with fine white edges to tertials only, notched blackish tail, clean white underparts, and indistinct thin white line from bill to top of eye. Tiny black feet. **HABITS AND HABITAT** Common resident of lowlands on both Caribbean and Pacific Slopes to 1,000m. Favours still or slow-flowing rivers, estuaries and some wetlands. Small groups skim low over water catching insects; often perches on dead trees, wires or overhanging branches near water. Out of breeding season gathers in large flocks. **SITES** Can be seen on boat trips on Rio Tárcoles and Caño Negro.

Grey-breasted Martin

■ *Progne chalybea* 17cm

DESCRIPTION The largest swallow in Costa Rica. Dark glossy blue-black upperparts extending over crown. Pale greyish underparts, palest on undertail, dark grey breast-band, paler grey throat, and dark blackish wings and deep, forked tail with slight blue gloss. Black bill and feet. **HABITS AND HABITAT** Fairly common resident throughout Costa Rica to 1,700m. Found in towns and open areas, and near bridges. Most often seen perched on wires and TV aerials around towns. Glides on flat wings, catching variety of insects while on the wing. Out of breeding season roosts in large numbers on wires or under bridges. **SITES** May be in any town, but always found around Puerto Viejo and La Selva OTS.

Rufous-backed Wren ■ *Campylorhynchus capistratus* 17cm

DESCRIPTION Large brown-and-white wren. Hind-neck and upper back rufous; rest of upperparts and wings dark brown with pale grey and black barring. Top of head dark brown. Fan-shaped tail also brown, with blackish barring and white tips. White underparts, including face, which has dark black line through eye and white brow over eye. Brown eyes, long, slightly curved dark bill and blue-grey legs. **HABITS AND HABITAT** Common resident of northern Pacific lowlands and western Central Valley to 1,000m. Favours open woodland, savannah, dry scrub and gardens. Usually in noisy family groups, staying low in search of insects such as beetles, and spiders. Utters various calls, but regular 'do-whip-de weeel'. **SITES** Carara NP, La Ensenada and Palo Verde.

Cabanis's Wren *Cantorchilus modestus* 13cm

DESCRIPTION Also includes newly split Canebrake and Isthmian Wren. Upperparts dull brown, more rufous on rump, tail and undertail, with buffy flanks. Fine blackish barring on wings and tail, brown crown, greyish-white throat, face and underparts, dark line through eye and pale whitish brow above eye. Bill has dark upper and pale greyish lower mandible. Grey legs. **HABITS AND HABITAT** Fairly common resident on both Caribbean and Pacific Slopes and Central Valley to 2,000m. Favours thickets and bushes, overgrown gardens and weedy fields, where it keeps well concealed. Very active seeking insects and spiders. Call 'chinchinchirree'. **SITES** Hotel Bougainvillea, Las Esquinas Rainforest Lodge and Carara NP.

House Wren *Troglodytes aedon* 10cm

DESCRIPTION Very small and the most likely wren to be seen around habitation. Overall muddy-brown, paler on underparts; fine black barring on wings and tail, and faint buffy line over eye. Short, sharp bill with upper mandible dark and lower mandible greyish. Brownish legs. **HABITS AND HABITAT** Common and widespread resident over much of Costa Rica from lowlands to 2,800m. Uncommon in north-west Pacific area. Found mostly around human habitation, gardens, and parks and nearby scrub. Hunts low and creeps around like a mouse, feeding on insects with cocked tail. Nests in and around houses, sheds and tangled roots. Song a bubbling, rattling series of loud notes. **SITES** Probably around any accommodation you stay at.

Grey-breasted Wood Wren ■ *Henicorhina leucophrys* 10cm

DESCRIPTION Tiny highland counterpart of **White-breasted Wood Wren** *H. leucosticta*. Rufous-brown back, rufous wings and tail finely barred with black; grey breast; lower belly and undertail washed rufous. Whitish throat with a few blackish streaks, black-and-white chequered face, black stripe through eye, white brow above eye and dark brown crown. Brown eyes, thin dark bill and dark legs. **HABITS AND HABITAT** Common highland resident over 800m and to treeline. Favours low scrub, thickets, forest tangles and bamboo, where it searches for insects among the leaf litter and stick piles. Song a loud and very musical 'wip-riddl-de-doowoo-wip'. **SITES** Monteverde, Savegre Lodge and Paraiso Quetzales.

Black-faced Solitaire ■ *Myadestes melanops* 17cm

DESCRIPTION Upperparts and underparts mostly slaty-blue-grey; wings blackish with all edges of feathers slaty-blue-grey. Black tail, jet-black face-mask, bright orange bill and orange legs. **HABITS AND HABITAT** Common resident of middle to high elevations at 1,000–2,500m. Favours wet montane forests with shady cover, bamboo and forest edges. Normally shy unless feeding on exposed berries, and difficult to locate as loud song is hard to pinpoint. Eats fruits and can associate with mixed species flocks. Metallic song sounds like squeaky gate, 'sweeeoo-sweeeoo-swee-ee'. **SITES** Monteverde, Savegre Mountain Lodge and Paraiso Quetzales.

Black-billed Nightingale-thrush ■ *Catharus gracilirostris* 15cm

DESCRIPTION Small, plump thrush associated with highlands. Upperparts dark olive-brown including wings and tail. Underparts slaty-grey; paler grey-white in centre on belly and throat. Olive-grey breast-band and flanks, slaty-grey head, dark eyes and thin black bill. The only Nightingale-thrush with entirely black bill. Brownish legs. **HABITS AND HABITAT** Common resident in southern highlands from 2,200m to timberline. Favours wet montane forests, especially oak forests, clearings, scrubby gardens and high paramo. Nearly always on or near the ground, where it hops and makes short runs in search of insects, often perching on low tree stumps. Also feeds on berries. Song a high-pitched, metallic whistle, 'pss-teweedewee'. **SITES** Savegre Mountain Lodge, Volcano Poás NP and Tapantí NP.

Orange-billed Nightingale-thrush ■ *Catharus aurantiirostris* 16cm

DESCRIPTION Small, plump thrush of middle elevations. Upperparts warm olive-brown, including wings, rump and tail. Underparts pale grey with white in centre of lower belly and small, pale grey patch on throat. Grey face, warm brown crown and back of neck, clean bright orange eye-ring surrounding dark eye, bright orange bill and pale orange legs. **HABITS AND HABITAT** Fairly common resident localized to middle elevation in southern and central Costa Rica, with an isolated population in the Nicoya Peninsula at 500–1,500m. Favours understorey of humid forests, thickets, coffee plantations and gardens. Forages low on the ground for insects, but will eat berries as well. Song a high-pitched, squeaky whistle, 'pt-sweeo-twee-ooo'. **SITES** Talari Mountain Lodge, El Pelicano Lodge and Tapantí NP.

Ruddy-capped Nightingale-thrush ■ *Catharus frantzii* 16cm

DESCRIPTION Small, plump thrush of the highlands. Upperparts, including back, rump, wings and tail, warm olive-brown. Upper back, nape and especially crown richer rufous-brown; underparts slaty-grey with paler grey centre to belly, undertail and throat. Face slaty-grey and bill with dark upper mandible and orange lower mandible. Brownish legs. **HABITS AND HABITAT** Common resident of highlands at 1,400–2,600m. Favours wet montane forest edges, clearings, thickets, bamboo and gardens. Forages on or near the ground in search of insects, and most often comes out into the open at dusk and dawn. Also feeds on berries. Song a slow, high-pitched, metallic 'sweeoodooly-sweee-sweeooly-sweedooly-dooly'. **SITES** Savegre Mountain Lodge and Paraiso Quetzales.

Swainson's Thrush ■ *Catharus ustulatus* 17cm

DESCRIPTION Small, plump thrush. Upperparts including wings and tail warm olive-brown; most of head and cheek warm brown. Throat and upper breast have pale orangey-yellow wash fading to white on belly and undertail. Heavy black spots from sides of throat, covering most of breast to mid-belly. Distinctive buffy-white eye-ring, and bill has dark upper mandible and pale pink-yellow lower mandible. Pinkish legs. **HABITS AND HABITAT** Common migrant found throughout Costa Rica in September–May to 2,800m. Because it is a migrant it may appear anywhere, although it favours wet forests, secondary growth and gardens, where it can often be found low down, feeding on berries and other fruits. Call a quick 'quip'. **SITES** Almost anywhere, especially around fruiting trees.

Wood Thrush ■ *Hylocichla mustelina* 19cm

DESCRIPTION Medium-sized, bright rufous thrush. Upperparts including top of head and neck, and wings bright rufous; rump and tail slightly darker olive-brown. Clean white underparts including throat and sides of face, with distinct black spotting, heaviest on breast and belly. White eye-ring, and mostly black bill with lower mandible paler pink-yellow at base. Pinkish legs. **HABITS AND HABITAT** Fairly common migrant in September–April in lowlands of both Caribbean and Pacific Slopes to 1,800m. Favours mature wet forests, thickets and damp secondary growth with leaf litter. Hops around the ground turning leaves in search of insects; also feeds on berries and other fruits. Call a loud 'pitt-pitt-pitt-pitt'. **SITES** Carara NP and La Selva OTS.

Sooty Thrush

■ *Turdus nigrescens* 25cm

DESCRIPTION Large, plump, all-black looking thrush. Entire upperparts, underparts, wings, tail and head sooty-black. Female slightly paler brown-black than male. Obvious pale whitish eye helps identify it, as do orange eye-ring, bill and legs. **HABITS AND HABITAT** Common resident of highlands above 2,000m to treeline. Can be found in open and semi-open areas, roadside verges, gardens, woodland edges and paramo. Hops around the ground turning leaf litter in search of insects and spiders, and also visits trees and bushes for berries and other fruits. Often sits on fence posts in misty mountains. Call a rattling 'greeek-greek-greek-greek-grek'. **SITES** Savegre Mountain Lodge, Trogon Lodge and Dantica Lodge.

Mountain Thrush ■ *Turdus plebejus* 25cm

DESCRIPTION Medium-sized, dark brown thrush. Upperparts, underparts, wings, tail and head dark olive-brown. Slightly paler on belly and throat, the latter having indistinct blackish streaks. Separated from most other thrushes in area by lack of distinguishing marks and notably all-black bill. Black legs. **HABITS AND HABITAT** Fairly common resident of montane habitat from treeline down to 1,200m. Prefers to stay in trees within wet, mossy, epiphyte-laden forests, forest edges, and occasionally secondary growth and gardens. Forages on the ground for insects, and more regularly feeds on fruiting trees and berries. Call a hard 'whip-pip-pip'. **SITES** Savegre Mountain Lodge, Trogon Lodge and Paraiso Quetzales.

Clay-coloured Thrush ■ *Turdus greyi* 24cm

DESCRIPTION Medium-sized, buffy-brown thrush. Upperparts, underparts, wings, head and tail warm buffy-brown. Paler greyish-brown on lower belly and undertail, and paler buff throat with indistinct dark streaking. Distinguished from most other Costa Rican thrushes by yellow-green bill and reddish eyes. Greyish-pink legs. **HABITS AND HABITAT** Common and widespread resident over much of Costa Rica to 2,500m. The national bird of Costa Rica. Can be found in pretty much any habitat, although it favours open agricultural land, gardens, parks and semi-open woodlands, as well as human habitation. Hops around the ground in search of insects and worms, and also eats fruits. Call a drawn-out 'peeuuuuuuuw'. **SITES** Can be seen almost anywhere.

American Dipper ▪ *Cinclus mexicanusi* 18cm

DESCRIPTION Plump and very rounded, all-grey bird. Upperparts slaty-grey, darker on wings and crown. Underparts, including rump, belly, breast and throat, paler grey. Short grey tail and indistinct whitish eye-ring. Black bill and dark eyes, with legs an obvious very pale pink. **HABITS AND HABITAT** Fairly common resident in right habitat, which is mostly mountain streams at 800–2,600m. Occurs almost exclusively at fast-flowing mountain streams with rocks and boulders, on which it perches, bobbing its body up and down. Dives under water to feed on aquatic insects and larvae. Call a sharp 'tsik'. **SITES** Savegre Mountain Lodge, Trogon Lodge and Bosque de Paz.

Yellow-throated Euphonia ▪ *Euphonia hirundinacea* 10cm

DESCRIPTION Tiny blue-and-yellow bird with short tail. Upperparts, including wings and tail, deep glossy blue. Underparts including throat bright yellow, except small white patch on belly. Glossy blue mask through face on mid-crown, nape and back of neck. Small patch of bright yellow on forehead. Pay particular attention to where this yellow reaches just above front of eye. This and yellow throat help to identify the species. Thickish, silvery-grey bill and grey legs. Female olive-green. **HABITS AND HABITAT** Fairly common to uncommon on Pacific Slope and in northern Caribbean area to 1,400m. Favours forest edges and secondary growth. Feeds mostly on mistletoe. Call a 'cheereee' or 'chee-cheet'. **SITES** Feeders at Arenal Observatory Lodge.

Olive-backed Euphonia ■ *Euphonia gouldi* 10cm

DESCRIPTION Rather plain-looking, all-green euphonia. Upperparts dark olive-green, including wings and short tail. Chest to upper belly olive-green; centre of lower belly and undertail rich chestnut. Head olive-green with male having patch of yellow on forecrown, and female patch of chestnut on forecrown. Bill black above and silver below, and greyish legs. **HABITS AND HABITAT** Common resident of Caribbean lowlands to 1,000m. Favours wet forests and forest edges, tall secondary growth and semi-open areas. Often in canopy, coming lower to feed on berries and other fruits, and mistletoe. Call 'prrrr-prrrr'; song 'pertooo-chip-cherry-weep'. **SITES** La Selva OTS, Rancho Naturalista and La Tirimbina Lodge.

Golden-browed Chlorophonia ■ *Chlorophonia callophrys* 13cm

DESCRIPTION Small plump and very bright green. On male, bright green upperparts including wings and tail; lower half of face, throat, upper breast and flanks also bright green. Rest of breast, belly and undertail bright yellow, with dark line on breast separating it from the green. Powder-blue line on sides and back of neck plus crown. Broad gold-yellow brow over eye. Greyish bill and legs. Female all green with yellow on belly and undertail. **HABITS AND HABITAT** Common resident of highlands above 1,000m to timberline. Often difficult to locate as it stays high in canopy of wet montane forests. Feeds on fruits and mistletoe. Call a mournful, pygmy-owl like, low, single note, 'poooo'. **SITES** Fairly reliably seen at Paraiso Quetzales.

Northern Waterthrush ■ *Parkesia noveboracensis* 14cm

DESCRIPTION Upperparts uniformly dark brown, including wings and tail. Underparts usually creamy-white with heavy black-brown streaks, including small streaks or flecks on throat and no streaking on undertail. Crown dark brown, face mottled grey-brown and long, thin, whitish stripe above eye. Bill black above and paler pink-brown below. Bright pink legs. **HABITS AND HABITAT** Common migrant in August–May throughout lowlands to 1,500m. Favours wet areas of forests, swamps, mangroves, ponds and stream edges, where it spends time walking on the ground, often bobbing its body. Call a hard, sharp 'tzik'. **SITES** Carara NP, La Selva OTS and Selva Verde.

Black-and-white Warbler ■ *Mniotilta varia* 13cm

DESCRIPTION The only black-and-white striped warbler found in Costa Rica. Head and back striped with black and white, underparts white streaked with black, wings black with two bold white wing-bars, and undertail white with black spots. Male has black throat, and female and juvenile have whitish pale throat. Dark greyish bill and brownish legs. **HABITS AND HABITAT** Common migrant in August–April throughout much of Costa Rica, mostly at 500–1,500m, although can occur both higher and lower. Can be found in any area of tall trees, where it is most often encountered with mixed species flocks. Climbs up and down trunks and along branches looking for small insects. Call an extremely high-pitched 'tsip'. **SITES** Found in mixed species flocks.

Prothonotary Warbler ■ *Protonotaria citrea* 13cm

DESCRIPTION Entire head and underparts very strikingly bright orange-yellow, back olive-green, wings and tail plain blue-grey, and a little white on belly. Bold black eyes stand out on clean face, and rather stout bill for a warbler is steel-blue-black. Blue-grey legs. **HABITS AND HABITAT** Locally common migrant in August–March, mostly in lowlands and along the coast to 1,500m. Preferred habitat is mangroves, river edges with thickets, and scrubby edges at streams and ponds. Stays low to the ground, and searches for insects among fallen trees and crevices. Call a rather soft 'chip'. **SITES** Best seen from boats on Rio Tárcoles.

Flame-throated Warbler ■ *Oreothlypis gutturalis* 12cm

DESCRIPTION Upperparts slaty-grey, including wings, rump and tail; black back. Head and face slaty-grey; throat and upper breast fiery red, more orange on female. Underparts greyish on flanks, and white down centre of belly and undertail. Thin black bill with yellow base to lower mandible, and brownish-grey legs. **HABITS AND HABITAT** Common resident of southern Central Highlands from 1,800m to timberline. Found mostly in trees and bushes in montane wet forests and forest edges, and semi-open habitat with scattered trees and bushes. Associates with mixed species flocks and rather acrobatic, hanging under leaves in search of small insects. Song a buzzy 'tt-tt-tt-bzzzzzzzzzzzzz'. **SITES** Savegre Mountain Lodge, Trogon Lodge and Paraiso Quetzales.

Tennessee Warbler ■ *Leiothlypis peregrina* 12cm

DESCRIPTION One of the plainest warblers, looking almost entirely grey-green. Upperparts grey-green to olive-green, yellowish on wing edges; underparts dull white, sometimes with yellowish-grey wash. Head of male greyer than female's, and both sexes have faint, thin whitish line over eye. Dark bill and grey legs. **HABITS AND HABITAT** Common migrant in September–May over much of Costa Rica to 2,400m. Favours semi-open and secondary growth, gardens, parks and in fact almost anywhere except extreme highlands. Often with mixed flocks of warblers, tanagers and honeycreepers, and feeds on insects, berries and sometimes other fruits. Call a sharp 'chip'. **SITES** Occurs among mixed species flocks; comes to fruit feeders at La Paz Waterfall Gardens.

Grey-crowned Yellowthroat ■ *Geothlypis poliocephala* 13cm

DESCRIPTION Upperparts bright green-olive, especially on wings and tail. Underparts bright yellow from throat right down to belly, lower belly and undertail slightly paler, and head green-olive with pale grey behind and above eye. Small, triangular black mask stretches from bill to eye, partial white eye-ring, bill dark above with paler pinkish lower mandible, and pinkish legs. Female's mask has less black than male's. **HABITS AND HABITAT** Common and widespread resident on both Caribbean and Pacific Slopes to 1,500m. Favours grassland and scrubby pastures where it stays low and hidden, occasionally coming to the top of a low bush to sing. Eats insects and some berries. Song 'wijewoo-wijewoo-wijiwijwiiiaa'. **SITES** Arenal area and La Selva OTS.

Tropical Parula ■ *Setophaga pitiayumi* 10cm

DESCRIPTION Very small. Upperparts deep blue, including head, wings and tail. Wings have one or sometimes two distinct pale whitish wing-bars; mossy-green back, yellow throat and belly, orange breast and white undertail. Black mask through eye, dark upper mandible and yellow lower mandible, and brown legs. **HABITS AND HABITAT** Fairly common resident of both Caribbean and Pacific Slopes at 600–1,800m. Prefers to stay in canopy of wet forests, forest edges, tall secondary growth and clearings. Often seen among mixed species flocks, which frequently include other warblers. Active as it seeks insects and small berries, and often hovers. Song a high-pitched trill, 'pppp-pseeeeeee-weeoow'. **SITES** Many places, but Arenal Observatory Lodge is a particularly good place to see it.

Blackburnian Warbler ■ *Setophaga fusca* 12cm

DESCRIPTION Overall this warbler looks black and white and streaky. Most birds visiting Costa Rica have only a warm orange-yellow throat, with just the odd few having the fabulous bright, fiery orange throat. Back black with white stripes, wing black with white patch, and tail black with white outer feathers. Underparts white with black streaks, black-and-white head pattern, and varying amount of orange on face and throat. Two-tone bill and grey legs. **HABITS AND HABITAT** Common migrant in August–April over most of Costa Rica at 500–1,500m. Found in most habitats with trees, and nearly always in mixed species flocks. Call a hard 'chip'. **SITES** May be found anywhere.

American Yellow Warbler ▪ *Setophaga petechia* 12cm

DESCRIPTION All-yellow warbler. Upperparts bright to dull yellow depending on age. Clean yellow head, wings with darker yellow-green centres to feathers, and yellow tail. Underparts clean pale yellow on many migrant birds; a few are brighter yellow, with faint to bold chestnut streaking. Black eyes and pinkish legs. The race *erithachorides*, found in mangroves, has chestnut head. **HABITS AND HABITAT** Common migrant throughout Costa Rica in August–May to 1,500m. Favours any wooded habitat, secondary growth and gardens. Resident race *erithachorides* confined mainly to Pacific mangroves. Active and restless as it forages for insects. Call a sharp 'chip'. **SITES** Migrants can occur anywhere. Residents easy to see from boats on Rio Tárcoles.

Chestnut-sided Warbler ▪ *Setophaga pensylvanica* 12cm

DESCRIPTION Most birds seen in Costa Rica are juveniles or in non-breeding plumage. Upperparts bright olive-green, including wings and tail, and upper half of head. Two bold creamy-white wing-bars. Grey face, neck, throat and underparts, and some birds have varying amount of chestnut on flanks. Bold white eye-ring and dark eyes. **HABITS AND HABITAT** Common migrant in most of Costa Rica in September–May to 2,000m. Found in all levels of forests, woodland, secondary growth and gardens, where it forages for insects. Habit of cocking tail up and drooping wings. Readily joins mixed species flocks and gives soft 'chip' call note. **SITES** Can occur anywhere, especially with flocks of other warblers.

Black-throated Green Warbler ■ *Setophaga virens* 12cm

DESCRIPTION Upperparts olive-green, including back, wings, crown and tail. Dusky-green wings with two bold whitish wing-bars; white underparts with dark streaking on flanks, sometimes with varying amount of black around throat and chin. Yellow face with indistinct yellow-green line on cheek. Small black eyes, dark bill and brownish-grey legs. **HABITS AND HABITAT** Common migrant in highlands in October–April, and at 800–3,000m. Prefers to stay in canopy of tall forest edges, secondary growth, semi-open areas and gardens with scattered trees. Forages among leaves for insects, and delivers soft 'tsit' call. **SITES** Savegre Mountain Lodge, Trogon Lodge and Monteverde.

Wilson's Warbler ■ *Cardellina pusilla* 12cm

DESCRIPTION Totally yellow-looking warbler that could be confused with American Yellow Warbler (see p. 131). Entire upperparts including tail olive-yellow; underparts bright yellow with no streaking. Yellow face with olive cheeks, and in juvenile and female an olive crown; male has clean black cap. Bill dark on upper mandible and pale on lower mandible. Pinkish-orange legs.
HABITS AND HABITAT Common migrant mostly to highlands in September–May above 900m to treeline. Found at all levels, from canopy of tall forests, forest edges and secondary growth, to shrubs and bushes in semi-open areas and gardens. Occurs among mixed species flocks, and actively forages for insects and spiders. Call a rather sharp 'chip'. **SITES** Savegre Mountain Lodge, Trogon Lodge and Paraiso Quetzales.

Collared Whitestart ■ *Myioborus torquatus* 12cm

DESCRIPTION Similar to **Slate-throated Whitestart** M. *miniatus*, but easily distinguished by bright yellow face and narrow, slaty-grey breast-band. Rest of upperparts slaty-blue-grey, including wings and tail, the latter with white outer feathers. Underparts bright yellow fading to pale yellow on undertail; rufous centre to crown bordered by thin black line. Dark eyes, and black bill and legs. **HABITS AND HABITAT** Common resident of southern and central highlands from 1,500m to timberline. Favours wet montane forests, forest edges, secondary growth and scrubby gardens. Feeds on insects from ground level to mid-level, where often seen in pairs or with mixed species flocks. Can be very confiding. Song a long series of whistles, 'wijiwijiweewijiwijwewewijiweewi-weiwijiweewijiwee'. **SITES** Savegre Mountain Lodge and Paraiso Quetzales.

Montezuma Oropendola ■ *Psarocolius montezuma* Male 50cm; female 38cm

DESCRIPTION Large and mostly maroon-plum, with colourful face and bill. Big difference between size of male and female. Upperparts, including back, wings, rump, breast, belly and undertail, rich maroon-plum. Dark tail with bright yellow outer feathers, and black head and neck. Bare skin of cheeks pale blue, pink wattle from base of lower mandible, pale blue skin around eye, and pinkish casque leading to long, pointed, tapering black bill with bright orange tip. Black legs. **HABITS AND HABITAT** Common resident mostly found in Caribbean lowlands and Central Valley to 1,500m. Often in flocks and nests in colonies. Feeds on small creatures and fruits. Call consists of gurgles and squeaks, 'ssssszzzzz-guggleguglegugle-tszzzeeeep-bahooo'. **SITES** Arenal Observatory Lodge and Rancho Naturalista.

Baltimore Oriole ■ *Icterus galbula* 19cm

DESCRIPTION Medium-sized orange-and-black oriole. Black head, neck, upper breast and back. Orange shoulder-stripe, broad white wing-bar and white edges to all wing feathers. Tail dark with orange outer feathers; underparts and rump bright orange. Black upper mandible and silver lower mandible. Blue-grey legs. Juvenile and female olive tinged with orange, two white wing-bars, and pale underneath with orange wash to chest. **HABITS AND HABITAT** Common migrant over most of Costa Rica in September–May to 2,000m. Prefers to sit in tops of trees of forest edges, semi-open areas, savannah and gardens. Often in small groups, and feeds on nectar flowers, fruits and insects. Call a 'chew-chew-chew'. **SITES** Feeders at Arenal Observatory Lodge and La Quinta Country Inn.

Black-cowled Oriole ■ *Icterus prosthemelas* 19cm

DESCRIPTION Elegant, slim, black-and-yellow oriole. Upperparts black, including wings and tail; bright yellow rump and shoulder-patch. Entire head down to lower chest black; belly and undertail bright yellow. Sharp bill has black upper mandible and silvery lower mandible. Black legs. **HABITS AND HABITAT** Fairly common resident throughout Caribbean lowlands to 1,200m. Rare in southern Pacific lowlands. Favours forest edges, tall secondary growth with palms, banana plantations, river edges and gardens. Feeds on insects, fruits including berries, and flowering trees, often hanging upside down. Call a sharp 'tweep'; song a liquid 'weetweet-cheereeep'. **SITES** Regular along approach road to La Selva OTS and La Quinta Country Inn.

Bronzed Cowbird *Molothrus aeneus* 20cm

DESCRIPTION Fairly stocky black bird with thick neck and shortish tail. Entire body black, but in good light shows glossy bronze-green. Wings and tail have more bluish gloss. Most distinctive features are bright red eyes and thick-based, conical, silvery-black bill. Black legs. Female duller than male and browner underneath. **HABITS AND HABITAT** Common and widespread resident over much of Costa Rica to 1,800m. Favours agricultural land, open areas and roadsides. Feeds in small groups on the ground, or gathers into huge flocks that eat grain, seeds and insects. Male can fluff his neck out in display. Call a very high, metallic 'tssweeeeeee'. **SITES** Farms, grain silos and roadsides.

Melodious Blackbird *Dives dives* 25cm

DESCRIPTION Entirely black all over with a glossy blue sheen. Fairly long, roundish tail, and small head with long, very sharp, pointed bill. Black eyes and legs. Female similar to male, but duller brownish-black. **HABITS AND HABITAT** Fairly common and now widespread resident over most of Costa Rica to 2,200m. First recorded in the country in 1987, and it is amazing how quickly it has expanded its range. Favours grassy fields, pastures, secondary growth, forest edges and gardens. Walks over the ground in search of insects and caterpillars; also takes nectar from flowers. Most often seen in pairs, and sings from tops of bushes and even rooftops, 'chrrrrr – werrcheer-weer-werrcheer-weer-weer'. **SITES** Bougainvillea Hotel, La Ensenada and La Quinta Country Inn.

Nicaraguan Grackle ■ *Quiscalus nicaraguensis* Male 30cm; female 25cm

DESCRIPTION Looks similar to Great-tailed Grackle (see below) but smaller and male appears blacker and with slightly less purplish gloss. Female different from Great-tailed female in that she is much paler brown below and has obvious paler stripe above eye. Male has twisted-looking tail, and both sexes have pale yellow eyes, dark bill and black legs. **HABITS AND HABITAT** Uncommon overall in Costa Rica, but locally fairly common along Rio Caño Negro. Prefers to be near sites with water such as river banks, marshes, wetlands and pastures. Mostly in small groups or pairs, and can be seen walking around in search of insects. Song 'sweeeoo-sweep-sweep sweep'. **SITES** Best seen on boat trips on Caño Negro.

Great-tailed Grackle ■ *Quiscalus mexicanus* Male 43cm; female 33cm

DESCRIPTION Long and slender black bird with very long tail. Upperparts, underparts and head appear black with glossy purple sheen. Wings and tail have greenish gloss. Tail has an unusual shape, looking like a twisted, half-opened fan in male. Pale whitish eyes, black bill and fairly long, blackish legs. Female has dull brown upperparts, wings and tail, and paler brown underparts with an even paler throat and line over eye. **HABITS AND HABITAT** Common and widespread resident to 2,000m. Rarely found in forests, preferring open habitat, especially near human habitation. Walks on the ground and feeds on fruits, seeds, grain, insects, small lizards and scraps. Call a sharp 'chuck'. **SITES** Can be seen almost everywhere.

Bananaquit ■ *Coereba flaveola* 9.5cm

DESCRIPTION Very small and chunky, with short tail and striped head. Upperparts, including wings and tail, olive-grey; wings have small white patch at bases of primaries. Dull white tips to tail. Yellow-green rump, bright yellow breast and belly, white undertail, grey throat, and blackish head with long white stripe over eye. Black bill slightly decurved with sharp point. Grey legs. **HABITS AND HABITAT** Common resident of lowlands on both Caribbean and Pacific Slopes to 1,500m. Favours forest edges, open woodland, plantations and gardens. Creeps through shrubs and flowers in search of insects, nectar and fruits. Regular at hummingbird feeders. Call a high 'tss-ss-ss-ss-ss-ss-ss-ss'. **SITES** Arenal Observatory Lodge, La Quinta Country Inn and La Selva OTS.

Rufous-collared Sparrow ■ *Zonotrichia capensis* 13cm

DESCRIPTION Adult readily distinguished by crest and rufous neck. Upperparts, wings and tail mostly brown with black streaking. Wings have two narrow white wing-bars. Underparts mostly dull white with buffy flanks and undertail. Black patches on sides of neck, whitish throat, grey-and-black striped head, bright chestnut collar, white eye-ring, pale brownish bill and pink legs. **HABITS AND HABITAT** Common and widespread resident of middle elevations and highlands from 600m to timberline. Found in non-forested areas, with a preference for urban areas, fields, semi-open habitat and gardens. Hops on the ground in search of seeds. Song a 'sweet-weet-tooo-tooo'. **SITES** Savegre Mountain Lodge, Trogon Lodge and Paraiso Quetzales.

Volcano Junco ■ *Junco vulcani* 16cm

DESCRIPTION Fairly large, plump sparrow. Brown back with black streaks, brown wings with small white spots, brown tail, and slaty-grey underparts with brownish wash on flanks. Paler grey throat, slaty-grey head with warm brown crown and cheeks, black between bill and eyes, bright yellow eyes, and pinkish bill and legs. **HABITS AND HABITAT** Fairly common resident of highlands and paramo, mostly above 2,800m. Favours open, grassy areas and edges of paramo, and low scrub around volcanos and mountain tops. Often in pairs, foraging on the ground for seeds and small insects. Call a very high-pitched 'tsip'; song a bubbly 'chee-weer-chewer-chee'. **SITES** Best site by radio masts at top of Cerro de la Muerte.

Stripe-headed Sparrow ■ *Peucaea ruficauda* 18cm

DESCRIPTION Large sparrow with black-and-white striped head. Upperparts a combination of buffy-brown feathers with dark centres, chestnut on shoulder and brown tail. Grey breast, and buffy-brown belly and undertail. Pure white throat and brow, grey collar, black face-mask, and black crown with fine white centre stripe. Brown eyes, and bill with black upper mandible and grey-pink lower mandible. Pinkish legs. **HABITS AND HABITAT** Common resident of dry northern Pacific lowlands to 1,000m. Favours low, dry scrub, savannah, secondary growth, cactus and edges of fields with hedgerows. Often near the ground or sits on fence wires. Feeds on seeds and insects. Call a sharp 'chip'; song a series of squeaky twitters. **SITES** Rio Tárcoles area, La Ensenada and Palo Verde.

Olive Sparrow ■ *Arremonops rufivirgatus* 14cm

DESCRIPTION Upperparts olive-green, including back, rump, wings and tail. Underparts greyish-buff with paler whitish belly. Pale grey throat, and buffy-grey head with black stripe through eye and black line along edge of crown, which shows a brown centre. Two-tone bill dark above and pale below. Pink-grey legs. **HABITS AND HABITAT** Common resident of north-west Pacific lowlands to 900m. Prefers to keep low down in dense understorey of dry forests, forest edges and secondary growth. Forages on the ground among thick tangles, and searches for insects and spiders in leaf litter. Call a high-pitched 'zip'; song an accelerating 'tsiip-tsip-tsip-tsip-tsip-tsipttttttttttttttttttttt'. **SITES** Palo Verde and Santa Rosa.

Black-striped Sparrow ■ *Arremonops conirostris* 14cm

DESCRIPTION Almost identical to Olive Sparrow (see above), but has grey head with no buff. Upperparts olive-green, including back, rump, wings and tail. Small line of yellow on edge of shoulder. Underparts greyish-buff with paler whitish belly. Pale grey throat, grey head with black stripe through eye and black line along edge of crown, which shows a brown centre. Two-tone bill dark above and pale below. Greyish legs. **HABITS AND HABITAT** Common resident in lowlands of Caribbean and South Pacific Slopes to 1,500m; different area than for Olive Sparrow. Favours forest edges, thick undergrowth, plantations and gardens. Hops around on the ground, feeding on insects and seeds. Song 'chweep-chweer', followed by long series of accelerating 'cheer-cheer-cheer-cheer-cher-cher-cher-ch-ch-ch-chchchchch'. **SITES** Arenal Observatory Lodge and Bosque del Rio Tigre.

Orange-billed Sparrow ■ *Arremon aurantiirostris* 15cm

DESCRIPTION Upperparts olive-green, duskier on wings and tail; greyish flanks, yellow shoulder-patch, white underparts and throat, and black breast-band. Black face; long white brow bordered by black line above and greyish crown-stripe. Bright orange bill and pinkish-brown legs. **HABITS AND HABITAT** Common resident in lowlands of Caribbean and South Pacific Slopes to 1,200m. Favours understorey of shady wet forests, forest edges and secondary growth, where it hops around on the ground scratching leaf litter or staying hidden in thickets. Feeds on insects, seeds and berries. Call a very high-pitched, short, sharp 'sip'; song a longer series of sharp whistles. **SITES** La Quinta Country Inn, La Selva OTS and Rancho Naturalista.

Chestnut-capped Brush Finch ■ *Arremon brunneinucha* 18cm

DESCRIPTION Upperparts dark olive-green, including wings and tail. White throat, centre of breast and belly; grey flanks and olive-green undertail. Thin black band across chest, black face and chestnut crown. Small white spot in front of eye, black bill and dark legs. **HABITS AND HABITAT** Common resident of middle elevation on both Caribbean and Pacific Slopes at 900–2,500m. Favours understorey of wet montane forests, secondary growth and shady tangles. Forages on the ground among leaf litter in search of insects, spiders and other small creatures. Call a high-pitched 'sip'; song a series of high-pitched whistles and notes. **SITES** Bosque de Paz and Arenal Observatory Lodge.

Large-footed Finch ■ *Pezopetes capitalis* 20cm

DESCRIPTION Dull, stocky and dark green finch. Upperparts dark olive-green, including wings. Long tail even darker with olive edges; underparts slightly paler olive-green. Dark slaty-grey head, black throat and face, and black lines on either side of crown running down to nape. Reddish eyes and black bill. Large feet and thick legs are dark brownish. **HABITS AND HABITAT** Common resident of southern Central Highlands above 2,000m to highest peaks and paramo. Favours dense understorey, bamboo thickets and forest floor, where it is most often located by the sound it makes by scratching and throwing leaves around. Feeds on insects, seeds and berries. Call a high 'zip-zip'; song a thrush-like melody. **SITES** Savegre Mountain Lodge, Trogon Lodge and Dantica Lodge.

Yellow-thighed Finch ■ *Pselliophorus tibialis* 18cm

DESCRIPTION All sooty-black looking finch. Upperparts, wings and longish tail all sooty-black, underparts sooty-black with dark olive tones, head dark slaty-black, and throat and crown matt black. Most obvious feature if seen is bright yellow thighs. Black bill and dark legs. **HABITS AND HABITAT** Common resident of central and southern highlands above 1,500m to timberline. Favours understorey, thickets, secondary growth, bamboo and gardens near montane forests. Mostly hops around on the ground or creeps through thickets, but can also work through trees and bushes. Follows mixed species flocks in search of insects and berries. Song a very fast, liquid series of squeaky notes. **SITES** Savegre Mountain Lodge and Paraiso Quetzales.

Common Bush Tanager ■ *Chlorospingus flavopectus* 13.5cm

DESCRIPTION Drab olive-green, sparrow-like bird. Upperparts olive-green, including wings and tail. Underparts grey-white with yellow-olive across chest, flanks and undertail. Greyish throat, brownish-grey head with slaty-grey mask; most useful identification feature is elongated white spot behind eye. Brown eyes, black bill and grey legs. **HABITS AND HABITAT** Common resident in central Costa Rica, of middle to high elevations at 500–2,200m. Favours wet epiphyte-laden forests, where it keeps to low understorey unless travelling with mixed species flocks, when it roams higher. Feeds on insects, spiders and small fruits. Call a sharp 'tzip'; song an incessant 'tsep-tsep-tsep-tsep-tsep'. **SITES** Paraiso Quetzales and Savegre Mountain Lodge.

Sooty-capped Bush Tanager ■ *Chlorospingus pileatus* 13.5cm

DESCRIPTION Similar to Common Bush Tanager (see above) except in head pattern. Upperparts olive-green, including wings and tail. Underparts grey-white with yellow-olive across chest, flanks and undertail. Greyish throat with indistinct dark malar stripe, and blackish head with long white stripe that appears broken above eye. Red eyes, black bill and grey legs. **HABITS AND HABITAT** Common resident in southern highlands above 2,000m. Favours wet, moss-laden mountain forests, where it keeps to low understorey unless travelling with mixed species flocks, when it roams higher. Feeds on insects, spiders and berries. Call a sharp 'tsip'; song a 'pit-pitcheeer-pitcherea-cherrea-chrrrrrrer-cher'. **SITES** Paraiso Quetzales and Savegre Mountain Lodge.

Grey-headed Tanager ■ *Eucometis penicillata* 18cm

DESCRIPTION Upperparts bright olive-green, including back, rump, wings and fairly long tail. Underparts bright yellow from lower throat to undertail. Grey head with paler grey-whitish throat and short, scruffy crest. Bright red eyes, blackish bill and pink-grey legs. **HABITS AND HABITAT** Fairly common resident of Pacific lowlands and northern Caribbean lowlands to 1,200m. Favours understorey of dark, damp forests and taller secondary growth. Regularly follows army ant swarms and monkeys in search of insects, but also feeds on fruits including berries. Often nervous and flies fast and low through forests, calling a harsh 'tzip-tzip'. **SITES** Regular sites include Carara National Park and El Pelicano Lodge.

Crimson-collared Tanager ■ *Ramphocelus sanguinolentus* 18.5cm

DESCRIPTION Stunning-looking large tanager with crimson-red collar extending over crown and across breast. Rump and undertail also crimson-red. Rest of plumage glossy black, including black face-mask. Brilliant red eyes, creamy-white bill and black legs. **HABITS AND HABITAT** Uncommon to common resident of Caribbean lowlands to 1,100m, crossing to Pacific Slope. Found in forest edges, secondary growth, semi-open areas and gardens. Usually in pairs, and can accompany other tanagers in search of insects and fruits; often visits lodge feeders. Call a thin 'tsweee' or 'weeet'. **SITES** Regular at feeders at La Quinta in Sarapiqui and Arenal Observatory Lodge.

Passerini's Tanager

■ *Ramphocelus passerinii* 16cm

DESCRIPTION Male upperparts, including head, upper back, wings and tail, are smooth velvety-black; underparts, including throat, breast, belly and undertail, also velvety-black. Lower back, rump and uppertail very bright scarlet. Silvery-blue bill and black legs. Male **Cherrie's Tanager** *R. costaricensis* is identical. Female olive-green with silvery bill. Female Cherrie's is olive with orange on chest and rump. **HABITS AND HABITAT** Common resident of Caribbean lowlands to 1,500m. Cherrie's Tanager only found in southern Pacific lowlands. Favours forest edges, open scrubby areas, thickets and gardens. Often moves around in noisy groups through undergrowth in search of berries and other fruits, and insects. Call heard when group foraging, 'krik-krik – krik-krik'. **SITES** La Selva OTS, La Quinta Country Inn and La Tirimbina Lodge.

Blue-grey Tanager ■ *Thraupis episcopus* 15cm

DESCRIPTION As its name suggests, a blue-grey tanager. Back dark blue-grey, wings and tail bright blue; underparts paler grey, along with throat and head. Dark eyes, bill with dark upper mandible and silvery lower mandible, and dark grey legs. Female greyer than male. **HABITS AND HABITAT** Common resident throughout Costa Rica to 2,000m. One of the most recognizable birds in the country as it can be seen around human habitation, towns and gardens, as well as in open woodland and humid forests. Feeds on all kinds of fruit, including berries, nectar and insects. Song sounds like someone playing with a child's squeaky toy. **SITES** Many places, including feeders at La Quinta Country Inn and Arenal Observatory Lodge.

Palm Tanager ■ *Thraupis palmarum* 15cm

DESCRIPTION Greyish-looking tanger. Upperparts, including back, rump, upper-wing, head and entire underparts, look a mid- to pale grey, but when seen in good light they show mossy-green tinge. Flight feathers on wing form triangular black patch that contrasts strongly with grey body. Black eyes, bill black above and silvery on lower mandible, and grey legs. **HABITS AND HABITAT** Common resident over much of Costa Rica to 1,500m except north-west, where it is rarer. Favours open areas, plantations, palms, parks and gardens. Mainly feeds on fruits, especially figs, but also insects and spiders. Song a long series of chatters and whistles. **SITES** Many places, including feeders at Arenal Observatory Lodge.

Emerald Tanager ■ *Tangara florida* 12cm

DESCRIPTION Strikingly bright, lime-green-yellow tanager with black-streaked back, black primaries and black tail. Distinctive square-shaped black mark on face and black around base of bill, which is also dark grey-black. Dark eyes and grey legs. Bird appears overall bright lime-green, but yellower on head, rump and belly. **HABITS AND HABITAT** Fairly common resident, although mostly seen in just ones or twos. Canopy species that often accompanies mixed flocks, found in Caribbean foothills at 300–1100m, in lowland and montane forests, and also in secondary growth. Call a weak 'tsip'. **SITES** Regularly sighted at Arenal Observatory Lodge and Braulio Carrillo.

Silver-throated Tanager ■ *Tangara icterocephala* 13cm

DESCRIPTION Bright yellow back with heavy black streaks, black wings and tail feathers with bright green edges; underparts bright yellow. Bright yellow head, silvery-white throat and sides of neck; thin black line separates throat from face. Dark eyes, blackish bill and grey legs. Female a duller version of male. **HABITS AND HABITAT** Common resident of middle elevation on Caribbean and Pacific Slopes at 600–2,000m. Favours middle level or canopy of humid forests, forest edges, secondary growth and gardens, where it forages through leaves in search of insects and berries. Quite acrobatic and often comes to birdfeeders. Call notes include 'fzit' and also buzzy 'chew'. **SITES** Many sites, including Arenal Observatory Lodge and Braulio Carrillo NP.

Bay-headed Tanager ■ *Tangara gyrola* 13cm

DESCRIPTION Bright grass-green back, wings and tail; bright blue rump and underparts, including throat, and green undertail and lower belly. Rusty-red thighs, and rich chestnut head with thin yellow hind-collar. Bill black above and greyish below, and greyish legs. Female duller than male. **HABITS AND HABITAT** Common resident of South Pacific Slope and Caribbean foothills at 100–1,500m. Favours middle levels of wet forests, forest edges, secondary growth, semi-open areas and gardens. Often seen among mixed species flocks, where it forages for insects and a wide variety of berries. Most often heard call is gentle 'tsip'. **SITES** Arenal Observatory Lodge and Braulio Carrillo.

Golden-hooded Tanager ■ *Tangara larvata* 13cm

DESCRIPTION Jet-black back, turquoise-blue shoulder and rump; blackish tail and wing feathers with pale yellow-green edges. Broad black chest-band, centre of belly white, and flanks and undertail greenish-blue. Gold head with black on chin, lores and around eyes, bluish cheeks, black bill and greyish legs. **HABITS AND HABITAT** Common resident of lowlands and foothills on both Caribbean and South Pacific Slopes to 1,600m. Favours canopy of forests, forest edges, secondary growth, semi-open areas and gardens. Often seen in mixed species flocks. Feeds on berries and insects; also comes to birdfeeders. Call a sharp, fast rattle, 'tk-tk-tk-tk-tk-tk-tk-tk-tk'. **SITES** Comes to feeders at La Quinta Country Inn and Arenal Observatory Lodge.

Spangle-cheeked Tanager ■ *Tangara dowii* 13cm

DESCRIPTION Upperparts mostly black with wings and tail edged with blue. Shoulder blue and rump pale blue-green; underparts, including breast, belly and undertail, cinnamon. Front of face and throat black, and dark blue-black crown. Cheeks and ear-coverts spangled with pale blue, and upper breast spotted with black. Bill dark above and pale below, and greyish legs. **HABITS AND HABITAT** Common resident of central and southern highlands at 1,000–3,000m. Favours montane forest and forest edges with trees covered in epiphytes and mosses; also gardens. Forages with mixed species flocks for insects, spiders and berries. Call a high, thin 'sip'. **SITES** Savegre Mountain Lodge and Tapantí National Park.

Blue Dacnis ■ *Dacnis cayana* 11cm

DESCRIPTION Small and mostly blue; sometimes mistaken for much larger **Turquoise Cotinga** *Cotinga ridgwayi*. Male has black back, wings and tail feathers black with bright blue edges; underparts and head bright blue, blackish throat and black line in front of eye (lores). Female mostly green with bluish on head. Thin bill dark above and pink below, and pinkish-grey legs. **HABITS AND HABITAT** Fairly common resident of Caribbean lowlands and South Pacific Slope to 1,200m. Favours forest edges and tall trees in gardens, where it often sits right out in the open. Follows mixed species flocks in search of insects and fruits. Call a very high-pitched, thin 'siip'. **SITES** Talari Mountain Lodge.

Flame-coloured Tanager ■ *Piranga bidentata* 18cm

DESCRIPTION Male overall bright orange-red; female yellow-green. Both sexes have the same markings, including black streaks on back, black tail with white outer tips, blackish wings with two bold white wing-bars and white tips to tertials. Upper mandible black and lower mandible grey. Grey legs. **HABITS AND HABITAT** Fairly common resident of highlands above 1,200m to timberline. Favours montane forests, forest edges, plantations and gardens. Can often be seen low on bushes, where it feeds on fruiting trees, berries and insects. Call a hard 'ker-dick'; song 'chewee-cheerup-cheerup-chewee-chew'. **SITES** Savegre Mountain Lodge, Paraiso Quetzales and Trogon Lodge.

Tooth-billed Tanager ■ *Piranga lutea* 18cm

DESCRIPTION Male dark dusky-red all over. Brighter red on throat and belly, and touches of grey on cheeks and sometimes on wing. Female plain olive-yellow all over with brighter belly and greyish wash on flanks. Both sexes have black upper mandible and silvery lower mandible. Dark grey legs. **HABITS AND HABITAT** Fairly common resident of middle to high elevations on both Caribbean and Pacific Slopes at 600–1,800m. Favours wet montane forests, forest edges, secondary growth and gardens. Feeds on berries and insects, and accompanies mixed species flocks. Call a hard 'prriip-prriip'. **SITES** Seen really well at Arenal Observatory Lodge feeders.

Female

Male

Red-throated Ant-Tanager ■ *Habia fuscicauda* 18cm

DESCRIPTION Male overall dark dusky-red that can look almost blackish. Underparts lighter, with bright red throat and inconspicuous bright red crown; face can look greyish. Female dusky-olive that is paler yellow-olive on belly and brighter yellow on throat. Black bill and dark legs. **HABITS AND HABITAT** Fairly common resident of Caribbean lowlands to 600m. Prefers to stay low in shady forest thickets and forest edges, often near streams, where noisy groups work through dense undergrowth. Constantly on the move in search of fruits and insects, and can follow ant swarms. Harsh, scolding call is 'craackk-craackk-craackk'. **SITES** Comes to feeders at La Quinta Country Inn, and reliable at La Selva OTS.

Shining Honeycreeper ■ *Cyanerpes lucidus* 10cm

DESCRIPTION Tiny and mostly deep blue. Wings and tail black with deep blue edges to all flight feathers. Back, underparts and head deep blue, paler on forehead. Black throat and line in front of eye (lores). Fairly long, thin, slightly decurved black bill. Most distinctive feature is bright yellow legs. Female greenish with tinges of blue on head and streaked underparts. **HABITS AND HABITAT** Fairly common resident in lowlands of Caribbean Slope and South Pacific Slope to 1,200m. Prefers to keep to canopy of forests, forest edges, semi-open areas and secondary growth. Often alone or in pairs, and feeds on insects, fruits including berries, and nectar. Call a high 'see-see-see-see'. **SITES** Reliable at La Selva OTS.

Red-legged Honeycreeper ■ *Cyanerpes cyaneus* 10.5cm

DESCRIPTION Tiny and mostly azure-blue. Black back, wings and tail, and azure shoulder-patch. Underparts and head completely azure-blue with paler turquoise-blue on crown. Fairly long, thin, slightly decurved black bill, and most notable feature is bright red legs. In flight shows bright yellow underwing. Female and non-breeding males mostly olive-grey-green. **HABITS AND HABITAT** Common resident all along Pacific Slope and lowlands of northern central Caribbean to 1,200m. Prefers to stay high in forest edges and secondary growth, but also comes lower into gardens. Feeds on insects, berries and nectar. Call a piercing 'tseet'. **SITES** Arenal Observatory Lodge and La Quinta Country Inn.

Green Honeycreeper ■ *Chlorophanes spiza* 13cm

DESCRIPTION Male bright sea-green all over, including upperparts, underparts, wings and tail. Slightly bluer on wing edges. Head has black crown extending down through face to a point, making it look like a triangle. Red eyes, and slightly curved bill black along top and yellow below. Greyish legs. Female all green without black on head. **HABITS AND HABITAT** Fairly common resident on Caribbean and South Pacific Slopes to 1,400m. Favours canopy of forest edges and clearings, coming down to shrubs and gardens. Moves with mixed species flocks, especially around fruiting trees. Eats insects, berries and other fruits, and nectar. Call a 'chip-chip-chip-chip'. **SITES** Arenal Observatory Lodge, La Selva OTS and La Quinta Country Inn.

Slaty Flowerpiercer ■ *Diglossa plumbea* 10cm

DESCRIPTION Very small, warbler-sized bird. Male completely slaty blue-grey, darker on wings and tail. Underparts paler grey on lower belly and undertail. Female olive-brown all over, with paler cream belly and indistinct fine streaks on upper breast. Bill long and thin, with unusual fine-hooked tip, mostly black with pinkish base to lower mandible. Brownish legs. **HABITS AND HABITAT** Common resident of Central Highlands above 1,200m to timberline. Favours flowering shrubs and epiphytes as it uses its hooked bill to pierce bases of flowers to extract nectar. Also feeds on small insects. Song a very high-pitched 'tsee-seeee-seeee-seee-see-tp-tp'. **SITES** Gardens of Savegre Mountain Lodge and Trogon Lodge.

Nicaraguan Seed Finch ■ *Oryzoborus nuttingi* 15cm

DESCRIPTION Male is large finch that is completely black with huge, bright pink bill making it unmistakable. Female dark brown-chestnut that is slightly paler on underparts and throat. She has dark blackish bill, and both sexes have black legs. **HABITS AND HABITAT** Uncommon and localized resident of Caribbean lowlands to 600m. Favours wet, grassy marshland with scattered bushes, hedgerows or forest edges. Feeds mostly on grass seeds and can be found with other species of finch. Song a rich series of rolling whistles and twitters. **SITES** Marshes near La Selva OTS and fairly easy to see along road to Caño Negro.

Yellow-faced Grassquit ■ *Tiaris olivacea* 10cm

DESCRIPTION Small greenish finch. Upperparts olive-green, including wings, tail and most of head. Female paler on underparts and with yellowish eye-stripe and throat. Male has black chest and face, grey underparts, bright yellow throat edged with black, bright yellow stripe over eye, and bright yellow eye-ring. Bill dark above and silvery below. Grey-pink legs. **HABITS AND HABITAT** Common resident of both Caribbean and Pacific lowlands to 2,200m. Favours grassy fields, pastures, gardens and roadsides, where it is often found in small groups foraging on the ground for seeds. Song a high trill that sounds like a grasshopper, 'tztztztztztztzttztztztztztztz'. **SITES** Any roadside grassy areas, including Arenal, Tapantí NP and El Pelicano Lodge.

Buff-throated Saltator

■ *Saltator maximus* 20cm

DESCRIPTION Upperparts, including tail, wings, and nape and rear of crown, olive-green; white chin and buffy throat surrounded by broad black band, and rest of underparts grey shading to buff undertail. Head slaty-grey with narrow white stripe over eye. Dark greyish-black bill and legs. **HABITS AND HABITAT** Common resident over much of Costa Rica to 1,200m; rarer in north-west. Favours secondary growth, gardens, forest edges, plantations and semi-open areas. Often follows mixed species flocks, and feeds on variety of berries and other fruits. Call a sharp 'pseeeet'; song 'pop-cherrie-cherrie'. **SITES** Comes to feeders at La Quinta Country Inn.

Greyish Saltator ■ *Saltator coerulescens* 20cm

DESCRIPTION Upperparts, including wings, tail and head, uniformly slate-grey. White throat and chin edged with black malar stripe. Upper breast greyish, shading to buff on belly and cinnamon on undertail. Face slaty-grey with obvious white line over eye, white lower eyelid, black bill and greyish legs. **HABITS AND HABITAT** Fairly common resident of Central Valley and lowlands of Caribbean Slope and Gulf of Nicoya to 1,850m. Favours semi-open areas, gardens, shady plantations and secondary growth. Often in pairs and on tops of bushes. Feeds on variety of fruits, flowers and buds. Song a whistled 'peet-choochoo-weet'. **SITES** Hotel Bougainvillea and La Quinta Country Inn.

Checklist of the Birds of Costa Rica

Sequence, nomenclature and taxonomy follow IOC World Bird Names (v. 9.1). www.worldbirdnames.org.

A Endemics
E Endemic to Costa Rica
Ep Endemic to Costa Rica and Panama
Ec Endemic to Cocos Island

B Abundance
C Common in preferred habitat
Fc Fairly common in preferred habitat
Un Uncommon: difficult to find in preferred habitat or casual stray
R Rare: localized and difficult to find

C Status
R Resident: species that occurs all year round and with evidence of breeding
Rc Resident or migrant breeding species on Cocos Island
I Introduced resident: species not native but regularly occurring and with evidence of breeding
M Migrant: only appears at certain times of year and may be resident part of year, for example in winter, or casual as in the case of several seabirds
V Vagrant: out of its normal range and not expected to be seen in Costa Rica, with very few records

D Birdlife International Global Status
LC Least Concern
NT Near Threatened
VU Vulnerable
EN Endangered
CR Critically Endangered

Note *Bird with an asterisk is considered by many other authorities to be a full species.

English Name	Scientific Name	A	B	C	D
Tinamous	**Tinamidae**				
Great Tinamou	*Tinamus major*		C	R	NT
Highland Tinamou	*Nothocercus bonapartei*		Un	R	LC
Little Tinamou	*Crypturellus soui*		Fc	R	LC
Slaty-breasted Tinamou	*Crypturellus boucardi*		Un	R	LC
Thicket Tinamou	*Crypturellus cinnamomeus*		Un	R	LC
Ducks, Geese and Swans	**Anatidae**				
White-faced Whistling Duck	*Dendrocygna viduata*		R	V	LC
Black-bellied Whistling Duck	*Dendrocygna autumnalis*		C	R	LC
Fulvous Whistling Duck	*Dendrocygna bicolor*		Un	R	LC
Comb Duck	*Sarkidiornis sylvicola*		R	V	LC

English Name	Scientific Name	A	B	C	D
Muscovy Duck	*Cairina moschata*		Fc	R	LC
American Wigeon	*Anas americana*		Un	M	LC
Mallard	*Anas platyrhynchos*		R	V	LC
Blue-winged Teal	*Anas discors*		C	M	LC
Cinnamon Teal	*Anas cyanoptera*		R	M	LC
Northern Shoveler	*Anas clypeata*		Un	M	LC
White-cheeked Pintail	*Anas bahamenis*		R	V	LC
Northern Pintail	*Anas acuta*		R	M	LC
Green-winged Teal	*Anas carolinensis*		Un	M	LC
Canvasback	*Aythya valisineria*		R	V	LC
Redhead	*Aythya americana*		R	V	LC
Ring-necked Duck	*Aythya collaris*		Un	M	LC
Greater Scaup	*Aythya marila*		R	V	LC
Lesser Scaup	*Aythya affinis*		Fc	M	LC
Hooded Merganser	*Lophodytes cucullatus*		R	V	LC
Masked Duck	*Nomonyx dominicus*		R	R	LC
Ruddy Duck	*Oxyura jamaicensis*		R	R?	LC
Chachalacas, Curassows and Guans	**Cracidae**				
Plain Chachalaca	*Ortalis vetula*		Fc	R	LC
Grey-headed Chachalaca	*Ortalis cinereiceps*		Fc	R	LC
Crested Guan	*Penelope purpurascens*		Fc	R	LC
Black Guan	*Chamaepetes unicolor*	Ep	Fc	R	NT
Great Curassow	*Crax rubra*		Un	R	VU
New World Quails	**Odontophoridae**				
Buffy-crowned Wood Partridge	*Dendrortyx leucophrys*		Un	R	LC
Spot-bellied Bobwhite	*Colinus leucopogon*		C	R	LC
Crested Bobwhite	*Colinus cristatus*		R	R	LC
Marbled Wood Quail	*Odontophorus gujanensis*		Un	R	NT
Black-eared Wood Quail	*Odontophorus melanotis*		R	R	LC
Black-breasted Wood Quail	*Odontophorus leucolaemus*	Ep	Fc	R	LC
Spotted Wood Quail	*Odontophorus guttatus*		Un	R	LC
Tawny-faced Quail	*Rhynchortyx cinctus*		R	R	LC
Albatrosses	**Diomedeidae**				
Waved Albatross	Phoebastria irrorata		R	V	CR
Petrels and Shearwaters	**Procellariidae**				
Black-capped Petrel	*Pterodroma hasitata*		R	V	EN
Galapagos Petrel	*Pterodroma phaeopygia*		R	V	CR
Tahiti Petrel	*Psudobulweria rostrata*		R	V	NT
Black Petrel (Parkinson's Petrel)	*Procellaria parkinsoni*		Un	M	VU
Cory's Shearwater	*Calonectris diomedea*		R	V	LC
Wedge-tailed Shearwater	*Puffinus pacificus*		Fc	R	LC
Sooty Shearwater	*Puffinus griseus*		R	V	NT
Short-tailed Shearwater	*Puffinus tenuirostris*		R	V	LC
Christmas Shearwater	*Puffinus nativitatis*		R	V	LC
Pink-footed Shearwater	*Puffinus creatopus*		R	R	VU
Audubon's Shearwater	*Puffinus lherminieri*		R	V	LC
Galapagos Shearwater	*Puffinus subalaris*		Un	R	LC
Black-vented Shearwater	*Puffinus opisthomelas*		R	V	NT
Austral Storm Petrels	**Oceanitidae**				
Wilson's Storm Petrel	*Oceanites oceanicus*		Un	M	LC
White-faced Storm Petrel	*Pelagodroma marina*		R	V	LC
Northern Storm Petrels	**Hydrobatidae**				
Least Storm Petrel	*Oceanodroma microsoma*		Un	R	LC
Wedge-rumped Storm Petrel	*Oceanodroma tethys*		Un	R	LC
Leach's Storm Petrel	*Oceanodroma leucorhoa*		Fc	R	LC
Band-rumped Storm Petrel	*Oceanodroma castro*		Un	R	LC
Markham's Storm Petrel	*Oceanodroma markhami*		R	V	LC
Black Storm Petrel	*Oceanodroma melania*		Fc	R	LC

English Name	Scientific Name	A	B	C	D
Grebes	**Podicipedidae**				
Least Grebe	*Tachybaptus dominicus*		Fc	R	LC
Pied-billed Grebe	*Podilymbus podiceps*		Un	R	LC
Black-necked Grebe	*Podiceps nigricollis*		R	V	LC
Tropicbirds	**Phaethontidae**				
White-tailed Tropicbird	*Phaethon lepturus*		R	V	LC
Red-billed Tropicbird	*Phaethon aethereus*		Un	R	LC
Red-tailed Tropicbird	*Phaethon rubricauda*		R	V	LC
Storks	**Ciconiidae**				
Maguari Stork	*Ciconia maguari*		R	V	LC
Wood Stork	*Mycteria americana*		C	R	LC
Jabiru	*Jabiru mycteria*		Un	R	LC
Ibises and Spoonbills	**Threskiornithidae**				
Green Ibis	*Mesembrinibis cayennensis*		Un	R	LC
American White Ibis	*Eudocimus albus*		C	R	LC
Glossy Ibis	*Plegadis falcinellus*		Un	R	LC
White-faced Ibis	*Plegadis chihi*		R	V	LC
Roseate Spoonbill	*Platalea ajaja*		Fc	R	LC
Herons and Bitterns	**Ardeidae**				
Rufescent Tiger Heron	*Tigrisoma lineatum*		R	R	LC
Fasciated Tiger Heron	*Tigrisoma fasciatum*		Un	R	LC
Bare-throated Tiger Heron	*Tigrisoma mexicanum*		Fc	R	LC
Agami Heron	*Agamia agami*		R	R	VU
Boat-billed Heron	*Cochlearius cochlearius*		Fc	R	LC
American Bittern	*Botaurus lentiginosus*		R	M	LC
Pinnated Bittern	*Botaurus pinnatus*		Un	R	LC
Least Bittern	*Ixobrychus exilis*		Un	R	LC
Black-crowned Night Heron	*Nycticorax nycticorax*		Fc	R	LC
Yellow-crowned Night Heron	*Nyctanassa violacea*		Fc	R	LC
Green Heron	*Butorides virescens*		C	R	LC
Striated Heron	*Butorides striata*		R	V	LC
Western Cattle Egret	*Bubulcus ibis*		C	R	LC
Great Blue Heron	*Ardea herodias*		Fc	M	LC
Great Egret	*Ardea alba*		C	R	LC
Reddish Egret	*Egretta rufescens*		R	M	NT
Tricolored Heron	*Egretta tricolor*		Fc	M	LC
Little Blue Heron	*Egretta caerulea*		C	M	LC
Snowy Egret	*Egretta thula*		C	R	LC
Pelicans	**Pelicanidae**				
American White Pelican	*Pelecanus erythrorhynchos*		R	V	LC
Brown Pelican	*Pelecanus occidentalis*		C	R	LC
Frigatebirds	**Fregatidae**				
Magnificent Frigatebird	*Fregata magnificens*		C	R	LC
Great Frigatebird	*Fregata minor*		C	Rc	LC
Gannets and Boobies	**Sulidae**				
Blue-footed Booby	*Sula nebouxii*		Un	M	LC
Masked Booby	*Sula dactylatra*		Un	M	LC
Nazca Booby	*Sula granti*		Un	M	LC
Red-footed Booby	*Sula sula*		Un	M	LC
Brown Booby	*Sula leucogaster*		C	R	LC
Cormorants and Shags	**Phalacrocoracidae**				
Neotropic Cormorant	*Phalacrocorax brasilianus*		C	R	LC
Anhingas and Darters	**Anhingidae**				
Anhinga	*Anhinga anhinga*		Fc	R	LC
New World Vultures	**Cathartidae**				
Turkey Vulture	*Cathartes aura*		C	R	LC
Lesser Yellow-headed Vulture	*Cathartes burrovianus*		Un	R	LC
Black Vulture	*Coragyps atratus*		C	R	LC

English Name	Scientific Name	A	B	C	D
King Vulture	*Sarcoramphus papa*		Un	R	LC
Ospreys	**Pandionidae**				
Western Osprey	*Pandion haliaetus*		C	M	LC
Kites, Hawks and Eagles	**Accipitridae**				
White-tailed Kite	*Elanus leucurus*		Fc	R	LC
Pearl Kite	*Gampsonyx swainsonii*		Un	R	LC
Grey-headed Kite	*Leptodon cayanensis*		Un	R	LC
Hook-billed Kite	*Chondrohierax uncinatus*		Un	R	LC
Swallow-tailed Kite	*Elanoides forficatus*		C	M	LC
Crested Eagle	*Morphnus guianensis*		R	R	NT
Harpy Eagle	*Harpia harpyja*		R	V	NT
Black Hawk-eagle	*Spizaetus tyrannus*		Un	R	LC
Black-and-white Hawk-Eagle	*Spizaetus melanoleucus*		R	R	LC
Ornate Hawk-eagle	*Spizaetus ornatus*		Un	R	NT
Double-toothed Kite	*Harpagus bidentatus*		Fc	R	LC
Tiny Hawk	*Accipiter superciliosus*		Un	R	LC
Sharp-shinned Hawk	*Accipiter striatus*		Un	M	LC
Cooper's Hawk	*Accipiter cooperii*		Un	M	LC
Bicoloured Hawk	*Accipiter bicolor*		R	R	LC
Northern Harrier	*Circus hudsonius*		R	M	LC
Grey-bellied Hawk	*Accipiter poliogaster*		R	V	NT
Mississippi Kite	*Ictinia mississippiensis*		Fc	M	LC
Plumbeous Kite	*Ictinia plumbea*		Fc	M	LC
Black-collared Hawk	*Busarellus nigricollis*		Un	R	LC
Snail Kite	*Rostrhamus sociabilis*		Fc	R	LC
Crane Hawk	*Geranospiza caerulescens*		Un	R	LC
Common Black Hawk	*Buteogallus anthracinus*		Fc	R	LC
Savanna Hawk	*Buteogallus meridionalis*		R	V	LC
Great Black Hawk	*Buteogallus urubitinga*		Un	R	LC
Solitary Eagle	*Buteogallus solitarius*		R	R	NT
Barred Hawk	*Morphnarchus princeps*		Fc	R	LC
Roadside Hawk	*Rupornis magnirostris*		C	R	LC
Harris's Hawk	*Parabuteo unicinctus*		Un	R	LC
White-tailed Hawk	*Geranoaetus albicaudatus*		Un	R	LC
White Hawk	*Pseudastur albicollis*		Fc	R	LC
Semiplumbeous Hawk	*Leucopternis semiplumbeus*		Un	R	LC
Grey Hawk	*Buteo plagiatus*		Fc	R	LC
Grey-lined Hawk	*Buteo nitidus*		Un	R	LC
Broad-winged Hawk	*Buteo platypterus*		C	M	LC
Short-tailed Hawk	*Buteo brachyurus*		C	R	LC
Swainson's Hawk	*Buteo swainsoni*		C	M	LC
Zone-tailed Hawk	*Buteo albonotatus*		Un	R	LC
Red-tailed Hawk	*Buteo jamaicensis*		Fc	R	LC
Sunbittern	**Eurypygidae**				
Sunbittern	*Eurypyga helias*		Un	R	LC
Finfoots	**Heliornithidae**				
Sungrebe	*Heliornis fulica*		Un	R	LC
Rails, Crakes and Coots	**Rallidae**				
Ocellated Crake	*Micropygia schomburgkii*		R	R	LC
Ruddy Crake	*Laterallus ruber*		R	V	LC
White-throated Crake	*Laterallus albigularis*		C	R	LC
Grey-breasted Crake	*Laterallus exilis*		Un	R	LC
Black Rail	*Laterallus jamaicensis*		R	R	NT
Clapper Rail	*Rallus longirostris*		R	R	LC
Rufous-necked Wood Rail	*Aramides axillaris*		R	R	LC
Grey-necked Wood Rail	*Aramides cajaneus*		C	R	LC
Uniform Crake	*Amaurolimnas concolor*		Un	R	LC
Sora	*Porzana carolina*		Fc	M	LC

English Name	Scientific Name	A	B	C	D
Yellow-breasted Crake	*Porzana flaviventer*		R	R	LC
Paint-billed Crake	*Neocrex erythrops*		R	R	LC
Spotted Rail	*Pardirallus maculatus*		R	R	LC
Purple Gallinule	*Porphyrio martinicus*		C	R	LC
Common Gallinule	*Gallinula galeata*		C	R	LC
American Coot	*Fulica americana*		Fc	R	LC
Limpkin	**Aramidae**				
Limpkin	*Aramus guarauna*		Fc	R	LC
Stone-curlews and Thick-knees	**Burhinidae**				
Double-striped Thick-knee	*Burhinus bistriatus*		Fc	R	LC
Oystercatchers	**Haematopodidae**				
American Oystercatcher	[illegible]		Un	M	LC
Stilts and Avocets	**Recurvirostridae**				
Black-necked Stilt	*Himantopus mexicanus*		C	R	LC
American Avocet	*Recurvirostra americana*		R	M	LC
Plovers	**Charadriidae**				
Southern Lapwing	*Vanellus chilensis*		Un	R	LC
Pacific Golden Plover	*Pluvialis fulva*		R	V	LC
American Golden Plover	*Pluvialis dominica*		R	M	LC
Grey Plover (Black-bellied Plover)	*Pluvialis squatarola*		C	M	LC
Semipalmated Plover	*Charadrius semipalmatus*		C	M	LC
Piping Plover	*Charadrius melodus*		R	V	NT
Wilson's Plover	*Charadrius wilsonia*		C	R	LC
Killdeer	*Charadrius vociferus*		Fc	M	LC
Snowy Plover	*Charadrius nivosus*		R	M	NT
Collared Plover	*Charadrius collaris*		Un	R	LC
Jacanas	**Jacanidae**				
Northern Jacana	*Jacana spinosa*		C	R	LC
Wattled Jacana	*Jacana jacana*		R	V	LC
Sandpipers and Snipes	**Scolopacidae**				
Wilson's Snipe	*Gallinago delicata*		Un	M	LC
Short-billed Dowitcher	*Limnodromus griseus*		C	M	LC
Long-billed Dowitcher	*Limnodromus scolopaceus*		R	M	LC
Hudsonian Godwit	*Limosa haemastica*		R	V	LC
Marbled Godwit	*Limosa fedoa*		Un	M	LC
Whimbrel	*Numenius phaeopus*		C	M	LC
Long-billed Curlew	*Numenius americanus*		R	M	LC
Upland Sandpiper	*Bartramia longicauda*		Un	M	LC
Greater Yellowlegs	*Tringa melanoleuca*		Un	M	LC
Lesser Yellowlegs	*Tringa flavipes*		C	M	LC
Solitary Sandpiper	*Tringa solitaria*		C	M	LC
Wandering Tattler	*Tringa incana*		Un	M	LC
Willet	*Tringa semipalmata*		C	M	LC
Spotted Sandpiper	*Actitis macularius*		C	M	LC
Ruddy Turnstone	*Arenaria interpres*		C	M	LC
Surfbird	*Aphriza virgata*		Un	M	LC
Red Knot	*Calidris canutus*		Un	M	LC
Sanderling	*Calidris alba*		Fc	M	LC
Semipalmated Sandpiper	*Calidris pusilla*		Fc	M	NT
Western Sandpiper	*Calidris mauri*		C	M	LC
Least Sandpiper	*Calidris minutilla*		C	M	LC
White-rumped Sandpiper	*Calidris fuscicollis*		R	M	LC
Baird's Sandpiper	*Calidris bairdii*		Un	M	LC
Pectoral Sandpiper	*Calidris melanotos*		Fc	M	LC
Dunlin	*Calidris alpina*		R	M	LC
Stilt Sandpiper	*Calidris himantopus*		Un	M	LC
Curlew Sandpiper	*Calidris ferruginea*		R	V	LC
Buff-breasted Sandpiper	*Tryngites subruficollis*		R	M	NT

English Name	Scientific Name	A	B	C	D
Ruff	*Philomachus pugnax*		R	V	LC
Wilson's Phalarope	*Phalaropus tricolor*		Fc	M	LC
Red-necked Phalarope	*Phalaropus lobatus*		Fc	M	LC
Red Phalarope	*Phalaropus fulicarius*		Un	M	LC
Gulls, Terns and Skimmers	**Laridae**				
Brown Noddy	*Anous stolidus*		Un	M	LC
Black Noddy	*Anous minutus*		C	Rc	LC
White Tern	*Gygis alba*		C	Rc	LC
Black Skimmer	*Rynchops niger*		Fc	M	LC
Swallow-tailed Gull	*Creagrus furcatus*		R	V	LC
Black-legged Kittiwake	*Rissa tridacyla*		R	V	LC
Sabine's Gull	*Xema sabini*		Fc	M	LC
Bonaparte's Gull	*Chroicocephalus philadelphia*		R	V	LC
Grey-headed Gull	*Chroicocephalus cirrocephalus*		R	V	LC
Laughing Gull	*Leucophaeus atricilla*		C	R	LC
Franklin's Gull	*Leucophaeus pipixcan*		Fc	M	LC
Grey Gull	*Leucophaeus modestus*		R	V	LC
Ring-billed Gull	*Larus delawarensis*		R	M	LC
California Gull	*Larus californicus*		R	V	LC
American Herring Gull	*Larus smithsonianus*		R	M	LC
Gull-billed Tern	*Gelochelidon nilotica*		Un	M	LC
Caspian Tern	*Hydroprogne caspia*		Un	R	LC
Royal Tern	*Thalasseus maximus*		C	R	LC
Elegant Tern	*Thalasseus elegans*		Un	M	NT
Cabot's Tern	*Thalasseus acuflavidus*		Fc	M	LC
Least Tern	*Sternula antillarum*		Un	R	LC
Large-billed Tern	*Phaetusa simplex*		R	V	LC
Bridled Tern	*Onychoprion anaethetus*		Fc	M	LC
Sooty Tern	*Onychoprion fuscatus*		Un	M	LC
Common Tern	*Sterna hirundo*		C	M	LC
Arctic Tern	*Sterna paradisaea*		R	V	LC
Forster's Tern	*Sterna forsteri*		Un	M	LC
Black Tern	*Chlidonias niger*		Fc	M	LC
Skuas	**Stercorariidae**				
South Polar Skua	*Stercorarius maccormicki*		R	M	LC
Pomarine Skua	*Stercorarius pomarinus*		Fc	M	LC
Parasitic Jaeger	*Stercorarius parasiticus*		Un	M	LC
Long-tailed Jaeger	*Stercorarius longicaudus*		R	V	LC
Pigeons and Doves	**Columbidae**				
Rock Dove	*Columba livia*		C	I	LC
White-crowned Pigeon	*Patagioenas leucocephala*		R	V	LC
Scaled Pigeon	*Patagioenas speciosa*		Fc	R	LC
Band-tailed Pigeon	*Patagioenas fasciata*		C	R	LC
Pale-vented Pigeon	*Patagioenas cayennensis*		C	R	LC
Red-billed Pigeon	*Patagioenas flavirostris*		Fc	R	LC
Ruddy Pigeon	*Patagioenas subvinacea*		Fc	R	VU
Short-billed Pigeon	*Patagioenas nigrirostris*		C	R	LC
Mourning Dove	*Zenaida macroura*		Fc	R	LC
White-winged Dove	*Zenaida asiatica*		C	R	LC
Inca Dove	*Columbina inca*		C	R	LC
Common Ground Dove	*Columbina passerina*		C	R	LC
Plain-breasted Ground Dove	*Columbina minuta*		Un	R	LC
Ruddy Ground Dove	*Columbina talpacoti*		C	R	LC
Blue Ground Dove	*Claravis pretiosa*		Un	R	LC
Maroon-chested Ground Dove	*Claravis mondetoura*		R	R	LC
White-tipped Dove	*Leptotila verreauxi*		C	R	LC
Grey-headed Dove	*Leptotila plumbeiceps*		Fc	R	LC
Grey-chested Dove	*Leptotila cassinii*		Fc	R	LC

English Name	Scientific Name	A	B	C	D
Purplish-backed Quail-Dove	*Geotrygon lawrencii*		Un	R	LC
Buff-fronted Quail-Dove	*Geotrygon costaricensis*	Ep	Un	R	LC
Olive-backed Quail-Dove	*Geotrygon veraguensis*		Un	R	LC
Chiriqui Quail-Dove	*Geotrygon chiriquensis*	Ep	Un	R	LC
Violaceous Quail-dove	*Geotrygon violacea*		Un	R	LC
Ruddy Quail-dove	*Geotrygon montana*		Fc	R	LC
Cuckoos	**Cuculidae**				
Greater Ani	*Crotophaga major*		R	V	LC
Smooth-billed Ani	*Crotophaga ani*		Fc	R	LC
Groove-billed Ani	*Crotophaga sulcirostris*		C	R	LC
Striped Cuckoo	*Tapera naevia*		Un	R	LC
Pheasant Cuckoo	*Dromococcyx phasianellus*		R	R	LC
Lesser Ground Cuckoo	*Morococcyx erythropygus*		Fc	R	LC
Rufous-vented Ground Cuckoo	*Neomorphus geoffroyi*		R	R	VU
Squirrel Cuckoo	*Piaya cayana*		C	R	LC
Yellow-billed Cuckoo	*Coccyzus americanus*		Un	M	LC
Mangrove Cuckoo	*Coccyzus minor*		Un	M	LC
Cocos Cuckoo	*Coccyzus ferrugineus*	Ec	C	Rc	VU
Black-billed Cuckoo	*Coccyzus erythropthalmus*		R	M	LC
Barn Owls	**Tytonidae**				
Western Barn Owl	*Tyto alba*		Fc	R	LC
Owls	**Strigidae**				
Pacific Screech Owl	*Megascops cooperi*		Fc	R	LC
Tropical Screech Owl	*Megascops choliba*		Fc	R	LC
Bare-shanked Screech Owl	*Megascops clarkii*		Fc	R	LC
Vermiculated Screech Owl	*Megascops vermiculatus*		Un	R	LC
Great Horned Owl	*Bubo virginianus*		R	R?	LC
Mottled Owl	*Strix virgata*		Fc	R	LC
Black-and-white Owl	*Strix nigrolineata*		Fc	R	LC
Crested Owl	*Lophostrix cristata*		Un	R	LC
Spectacled Owl	*Pulsatrix perspicillata*		Fc	R	LC
Costa Rican Pygmy Owl	*Glaucidium costaricanum*	Ep	Un	R	LC
Central American Pygmy Owl	*Glaucidium griseiceps*		Un	R	LC
Ferruginous Pygmy Owl	*Glaucidium brasilianum*		C	R	LC
Burrowing Owl	*Athene cunicularia*		R	V	LC
Unspotted Saw-whet Owl	*Aegolius ridgwayi*		R	R	LC
Striped Owl	*Pseudoscops clamator*		Fc	R	LC
Short-eared Owl	*Asio flammeus*		R	V	LC
Oilbird	**Steatornithidae**				
Oilbird	*Steatornis caripensis*		R	R?	LC
Potoos	**Nyctibiidae**				
Great Potoo	*Nyctibius grandis*		Un	R	LC
Northern Potoo	*Nyctibius jamaicensis*		Un	R	LC
Common Potoo	*Nyctibius griseus*		Fc	R	LC
Nightjars	**Caprimulgidae**				
Lesser Nighthawk	*Chordeiles acutipennis*		Fc	R	LC
Common Nighthawk	*Chordeiles minor*		C	M	LC
Short-tailed Nighthawk	*Lurocalis semitorquatus*		Un	R	LC
Pauraque	*Nyctidromus albicollis*		C	R	LC
White-tailed Nightjar	*Hydropsalis cayennensis*		Un	R	LC
Ocellated Poorwill	*Nyctiphrynus ocellatus*		R	R?	LC
Chuck-will's-Widow	*Antrostomus carolinensis*		Un	M	LC
Rufous Nightjar	*Antrostomus rufus*		R	R	LC
Eastern Whip-poor-Will	*Antrostomus vociferus*		R	M	LC
Dusky Nightjar	*Antrostomus saturatus*	Ep	Fc	R	LC
Swifts	**Apodidae**				
Spot-fronted Swift	*Cypseloides cherriei*		Un	R	LC

English Name	Scientific Name	A	B	C	D
White-chinned Swift	*Cypseloides cryptus*		R	R	LC
American Black Swift	*Cypseloides niger*		Un	R	LC
Chestnut-collared Swift	*Streptoprocne rutila*		Fc	R	LC
White-collared Swift	*Streptoprocne zonaris*		C	R	LC
Costa Rican Swift	*Chaetura fumosa*	Ep	C	R	LC
Grey-rumped Swift	*Chaetura cinereiventris*		C	R	LC
Vaux's Swift	*Chaetura vauxi*		Fc	R	LC
Chimney Swift	*Chaetura pelagica*		C	M	NT
Great Swallow-tailed Swift	*Panyptila sanctihieronymi*		R	V	LC
Lesser Swallow-tailed Swift	*Panyptila cayennensis*		Fc	R	LC
Hummingbirds	**Trochilidae**				
White-tipped Sicklebill	*Eutoxeres aquila*		Un	R	LC
Bronzy Hermit	*Glaucis aeneus*		Un	R	LC
Band-tailed Barbthroat	*Threnetes ruckeri*		Fc	R	LC
Green Hermit	*Phaethornis guy*		C	R	LC
Long-billed Hermit	*Phaethornis longirostris*		C	R	LC
Stripe-throated Hermit	*Phaethornis striigularis*		C	R	LC
Green-fronted Lancebill	*Doryfera ludovicae*		Un	R	LC
Scaly-breasted Hummingbird	*Phaeochroa cuvierii*		C	R	LC
Violet Sabrewing	*Campylopterus hemileucurus*		C	R	LC
White-necked Jacobin	*Florisuga mellivora*		Fc	R	LC
Brown Violetear	*Colibri delphinae*		Un	R	LC
Lesser Violetear	*Colibri cyanotus*		C	R	LC
Green-breasted Mango	*Anthracothorax prevostii*		Fc	R	LC
Violet-headed Hummingbird	*Klais guimeti*		Fc	R	LC
Rufous-crested Coquette	*Lophornis delattrei*		R	V	LC
Black-crested Coquette	*Lophornis helenae*		Un	R	LC
White-crested Coquette	*Lophornis adorabilis*	Ep	Un	R	LC
Green Thorntail	*Discosura conversii*		Fc	R	LC
Canivet's Emerald	*Chlorostilbon canivetii*		C	R	LC
Garden Emerald	*Chlorostilbon assimilis*	Ep	Fc	R	LC
Fiery-throated Hummingbird	*Panterpe insignis*	Ep	C	R	LC
White-tailed Emerald	*Elvira chionura*		Un	R	LC
Coppery-headed Emerald	*Elvira cupreiceps*	E	Fc	R	LC
Stripe-tailed Hummingbird	*Eupherusa eximia*		Un	R	LC
Black-bellied Hummingbird	*Eupherusa nigriventris*	Ep	Un	R	LC
Crowned Woodnymph	*Thalurania colombica*		C	R	LC
Cinnamon Hummingbird	*Amazilia rutila*		Fc	R	LC
Sapphire-throated Hummingbird	*Lepidopyga coeruleogularis*		R	R?	LC
Blue-throated Sapphire	*Hylocharis eliciae*		Un	R	LC
Rufous-tailed Hummingbird	*Amazilia tzacatl*		C	R	LC
White-bellied Emerald	*Amazilia candida*		R	V	LC
Blue-chested Hummingbird	*Amazilia amabilis*		Un	R	LC
Charming Hummingbird	*Amazilia decora*	Ep	Fc	R	LC
Mangrove Hummingbird	*Amazilia boucardi*	E	Un	R	EN
Steely-vented Hummingbird	*Amazilia saucerottei*		Fc	R	LC
Snowy-bellied Hummingbird	*Amazilia edward*		Un	R	LC
Blue-tailed Hummingbird	*Amazilia cyanura*		R	R?	LC
Snowcap	*Microchera albocoronata*		Un	R	LC
Bronze-tailed Plumeleteer	*Chalybura urochrysia*		Un	R	LC
White-bellied Mountaingem	*Lampornis hemileucus*		Un	R	LC
Purple-throated Mountaingem	*Lampornis calolaemus*		Fc	R	LC
Grey-tailed Mountaingem	*Lampornis cinereicauda*	E	Fc	R	LC
Green-crowned Brilliant	*Heliodoxa jacula*		C	R	LC
Talamanca Hummingbird	*Eugenes spectabilis*		C	R	LC
Purple-crowned Fairy	*Heliothryx barroti*		Fc	R	LC
Plain-capped Starthroat	*Heliomaster constantii*		Un	R	LC
Long-billed Starthroat	*Heliomaster longirostris*		Un	R	LC

English Name	Scientific Name	A	B	C	D
Magenta-throated Woodstar	*Calliphlox bryantae*	Ep	Un	R	LC
Ruby-throated Hummingbird	*Archilochus colubris*		Fc	M	LC
Volcano Hummingbird	*Selasphorus flammula*	Ep	Fc	R	LC
Scintillant Hummingbird	*Selasphorus scintilla*	Ep	Fc	R	LC
Trogons	**Trogonidae**				
Resplendent Quetzal	*Pharomachrus mocinno*		Fc	R	NT
Lattice-tailed Trogon	*Trogon clathratus*	Ep	Un	R	LC
Slaty-tailed Trogon	*Trogon massena*		C	R	LC
Black-headed Trogon	*Trogon melanocephalus*		C	R	LC
Baird's Trogon	*Trogon bairdii*	Ep	Fc	R	NT
Gartered Trogon	*Trogon caligatus*		C	R	LC
Black-throated Trogon	*Trogon rufus*		C	R	LC
Elegant Trogon	*Trogon elegans*		Un	R	LC
Collared Trogon	*Trogon collaris*		Fc	R	LC
*Orange-bellied Trogon	*Trogon collaris aurantiiventris*	Ep	Fc	R	LC
Kingfishers	**Alcedinidae**				
American Pygmy Kingfisher	*Chloroceryle aenea*		Un	R	LC
Green-and-rufous Kingfisher	*Chloroceryle inda*		R	R	LC
Green Kingfisher	*Chloroceryle americana*		C	R	LC
Amazon Kingfisher	*Chloroceryle amazona*		Fc	R	LC
Ringed Kingfisher	*Megaceryle torquata*		Fc	R	LC
Belted Kingfisher	*Megaceryle alcyon*		Fc	M	LC
Motmots					
Tody Motmot	*Hylomanes momotula*		Un	R	LC
Lesson's Motmot	*Momotus lessonii*		C	R	LC
Rufous Motmot	*Baryphthengus martii*		Fc	R	LC
Keel-billed Motmot	*Electron carinatum*		R	R	VU
Broad-billed Motmot	*Electron platyrhynchum*		C	R	LC
Turquoise-browed Motmot	*Eumomota superciliosa*		C	R	LC
Jacamars	**Galbulidae**				
Rufous-tailed Jacamar	*Galbula ruficauda*		Fc	R	LC
Great Jacamar	*Jacamerops aureus*		R	R	LC
Puffbirds	**Bucconidae**				
White-necked Puffbird	*Notharchus hyperrhynchus*		Un	R	LC
Pied Puffbird	*Notharchus tectus*		Un	R	LC
White-whiskered Puffbird	*Malacoptila panamensis*		Fc	R	LC
Lanceolated Monklet	*Micromonacha lanceolata*		R	R	LC
White-fronted Nunbird	*Monasa morphoeus*		Un	R	LC
New World Barbets	**Capitonidae**				
Red-headed Barbet	*Eubucco bourcierii*		Un	R	LC
Toucan Barbets	**Semnornithidae**				
Prong-billed Barbet	*Semnornis frantzii*	Ep	Fc	R	LC
Toucans	**Ramphastidae**				
Blue-throated Toucanet	*Aulacorhynchus caeruleogularis*	Ep	Fc	R	LC
Collared Aracari	*Pteroglossus torquatus*		C	R	LC
Fiery-billed Aracari	*Pteroglossus frantzii*	Ep	Fc	R	LC
Yellow-eared Toucanet	*Selenidera spectabilis*		R	R	LC
Keel-billed Toucan	*Ramphastos sulfuratus*		C	R	LC
Yellow-throated Toucan	*Ramphastos ambiguus*		C	R	NT
Woodpeckers	**Picidae**				
Olivaceous Piculet	*Picumnus olivaceus*		Fc	R	LC
Acorn Woodpecker	*Melanerpes formicivorus*		C	R	LC
Golden-naped Woodpecker	*Melanerpes chrysauchen*	Ep	Un	R	LC
Black-cheeked Woodpecker	*Melanerpes pucherani*		C	R	LC
Red-crowned Woodpecker	*Melanerpes rubricapillus*		Fc	R	LC
Hoffmann's Woodpecker	*Melanerpes hoffmannii*		C	R	LC
Yellow-bellied Sapsucker	*Sphyrapicus varius*		Un	M	LC
Hairy Woodpecker	*Picoides villosus*		Fc	R	LC

English Name	Scientific Name	A	B	C	D
Smoky-brown Woodpecker	*Picoides fumigatus*		Un	R	LC
Red-rumped Woodpecker	*Veniliornis kirkii*		R	R	LC
Rufous-winged Woodpecker	*Piculus simplex*		Un	R	LC
Golden-olive Woodpecker	*Colaptes rubiginosus*		Fc	R	LC
Cinnamon Woodpecker	*Celeus loricatus*		Fc	R	LC
Chestnut-coloured Woodpecker	*Celeus castaneus*		Fc	R	LC
Lineated Woodpecker	*Dryocopus lineatus*		Fc	R	LC
Pale-billed Woodpecker	*Campephilus guatemalensis*		Fc	R	LC
Caracaras and Falcons	**Falconidae**				
Red-throated Caracara	*Ibycter americanus*		R	R	LC
Northern Crested Caracara	*Caracara cheriway*		C	R	LC
Yellow-headed Caracara	*Milvago chimachima*		C	R	LC
Laughing Falcon	*Herpetotheres cachinnans*		Fc	R	LC
Barred Forest Falcon	*Micrastur ruficollis*		Un	R	LC
Slaty-backed Forest Falcon	*Micrastur mirandollei*		R	R	LC
Collared Forest Falcon	*Micrastur semitorquatus*		Fc	R	LC
American Kestrel	*Falco sparverius*		Un	M	LC
Aplomado Falcon	*Falco femoralis*		R	V	LC
Merlin	*Falco columbarius*		Un	M	LC
Bat Falcon	*Falco rufigularis*		Un	R	LC
Orange-breasted Falcon	*Falco deiroleucus*		R	V	NT
Peregrine Falcon	*Falco peregrinus*		Fc	R	LC
Parrots	**Psittacidae**				
Great Green Macaw	*Ara ambiguus*		Un	R	EN
Scarlet Macaw	*Ara macao*		Fc	R	LC
Olive-throated Parakeet	*Eupsittula nana*		Fc	R	LC
Orange-fronted Parakeet	*Eupsittula canicularis*		Fc	R	LC
Brown-throated Parakeet	*Eupsittula pertinax*		Fc	R	LC
Finsch's Parakeet	*Psittacara finschi*		C	R	LC
Sulphur-winged Parakeet	*Pyrrhura hoffmanni*	Ep	Un	R	LC
Barred Parakeet	*Bolborhynchus lineola*		Un	R	LC
Orange-chinned Parakeet	*Brotogeris jugularis*		C	R	LC
Red-fronted Parrotlet	*Touit costaricensis*	Ep	R	R	VU
Brown-hooded Parrot	*Pyrilia haematotis*		Fc	R	LC
Blue-headed Parrot	*Pionus menstruus*		Fc	R	LC
White-crowned Parrot	*Pionus senilis*		C	R	LC
White-fronted Amazon	*Amazona albifrons*		Fc	R	LC
Red-lored Amazon	*Amazona autumnalis*		C	R	LC
Yellow-naped Amazon	*Amazona auropalliata*		Fc	R	VU
Northern Mealy Amazon	*Amazona guatemalae*		C	R	NT
Ovenbirds	**Furnariidae**				
Pale-breasted Spinetail	*Synallaxis albescens*		Fc	R	LC
Slaty Spinetail	*Synallaxis brachyura*		C	R	LC
Red-faced Spinetail	*Cranioleuca erythrops*		Fc	R	LC
Spotted Barbtail	*Premnoplex brunnescens*		C	R	LC
Ruddy Treerunner	*Margarornis rubiginosus*	Ep	C	R	LC
Buffy Tuftedcheek	*Pseudocolaptes lawrencii*		Un	R	LC
Scaly-throated Foliage-gleaner	*Anabacerthia variegaticeps*		R	R	LC
Lineated Foliage-gleaner	*Syndactyla subalaris*		Un	R	LC
Western Woodhaunter	*Hyloctistes virgatus*		Un	R	LC
Buff-fronted Foliage-gleaner	*Philydor rufum*		Un	R	LC
Streak-breasted Treehunter	*Thripadectes rufobrunneus*	Ep	Fc	R	LC
Buff-throated Foliage-gleaner	*Automolus ochrolaemus*		Fc	R	LC
Ruddy Foliage-gleaner	*Automolus rubiginosus*		R	R	LC
Tawny-throated Leaftosser	*Sclerurus mexicanus*		Un	R	LC
Grey-throated Leaftosser	*Sclerurus albigularis*		Un	R	NT
Scaly-throated Leaftosser	*Sclerurus guatemalensis*		Un	R	LC
Plain Xenops	*Xenops minutus*		Fc	R	LC

English Name	Scientific Name	A	B	C	D
Streaked Xenops	*Xenops rutilans*		Un	R	LC
Plain-brown Woodcreeper	*Dendrocincla fuliginosa*		Un	R	LC
Tawny-winged Woodcreeper	*Dendrocincla anabatina*		Fc	R	LC
Ruddy Woodcreeper	*Dendrocincla homochroa*		Un	R	LC
Long-tailed Woodcreeper	*Deconychura longicauda*		Un	R	NT
Olivaceous Woodcreeper	*Sittasomus griseicapillus*		Fc	R	LC
Wedge-billed Woodcreeper	*Glyphorynchus spirurus*		C	R	LC
Strong-billed Woodcreeper	*Xiphocolaptes promeropirhynchus*		R	R	LC
Northern Barred Woodcreeper	*Dendrocolaptes sanctithomae*		Fc	R	LC
Black-banded Woodcreeper	*Dendrocolaptes picumnus*		R	R	LC
Cocoa Woodcreeper	*Xiphorhynchus susurrans*		C	R	LC
Ivory-billed Woodcreeper	*Xiphorhynchus flavigaster*		[illegible]	R	LC
Black-striped Woodcreeper	*Xiphorhynchus lachrymosus*		Un	R	LC
Spotted Woodcreeper	*Xiphorhynchus erythropygius*		C	R	LC
Streak-headed Woodcreeper	*Lepidocolaptes souleyetii*		C	R	LC
Spot-crowned Woodcreeper	*Lepidocolaptes affinis*		Fc	R	LC
Brown-billed Scythebill	*Campylorhamphus pusillus*		R	R	LC
Antbirds	**Thamnophilidae**				
Fasciated Antshrike	*Cymbilaimus lineatus*		Fc	R	LC
Great Antshrike	*Taraba major*		Fc	R	LC
Barred Antshrike	*Thamnophilus doliatus*		C	R	LC
Black-hooded Antshrike	*Thamnophilus bridgesi*	Ep	C	R	LC
Black-crowned Antshrike	*Thamnophilus atrinucha*		Un	R	LC
Russet Antshrike	*Thamnistes anabatinus*		Fc	R	LC
Plain Antvireo	*Dysithamnus mentalis*		Fc	R	LC
Streak-crowned Antvireo	*Dysithamnus striaticeps*		Un	R	LC
Spot-crowned Antvireo	*Dysithamnus puncticeps*		Un	R	LC
Checker-throated Antwren	*Epinecrophylla fulviventris*		Un	R	LC
White-flanked Antwren	*Myrmotherula axillaris*		Un	R	LC
Slaty Antwren	*Myrmotherula schisticolor*		Fc	R	LC
Dot-winged Antwren	*Microrhopias quixensis*		C	R	LC
Rufous-rumped Antwren	*Terenura callinota*		R	R	LC
Dusky Antbird	*Cercomacra tyrannina*		C	R	LC
Bare-crowned Antbird	*Gymnocichla nudiceps*		Un	R	LC
Chestnut-backed Antbird	*Myrmeciza exsul*		C	R	LC
Dull-mantled Antbird	*Myrmeciza laemosticta*		Fc	R	LC
Zeledon's Antbird	*Myrmeciza zeledoni*		Un	R	LC
Bicoloured Antbird	*Gymnopithys bicolor*		Un	R	LC
Spotted Antbird	*Hylophylax naevioides*		Fc	R	LC
Ocellated Antbird	*Phaenostictus mcleannani*		Un	R	LC
Antthrushes	**Formicariidae**				
Black-faced Antthrush	*Formicarius analis*		Fc	R	LC
Black-headed Antthrush	*Formicarius nigricapillus*		Un	R	LC
Rufous-breasted Antthrush	*Formicarius rufipectus*		Un	R	LC
Antpittas	**Grallariidae**				
Scaled Antpitta	*Grallaria guatimalensis*		R	R	LC
Streak-chested Antpitta	*Hylopezus perspicillatus*		Fc	R	LC
Thicket Antpitta	*Hylopezus dives*		Fc	R	LC
Ochre-breasted Antpitta	*Grallaricula flavirostris*		R	R	NT
Black-crowned Antpitta	*Pittasoma michleri*		R	R	LC
Tapaculos	**Rhinocryptidae**				
Silvery-fronted Tapaculo	*Scytalopus argentifrons*	Ep	Fc	R	LC
Tyrant Flycatchers	**Tyrannidae**				
Grey-headed Piprites	*Piprites griseiceps*		R	R	LC
White-fronted Tyrannulet	*Phyllomyias zeledoni*		R	R	LC
Yellow-crowned Tyrannulet	*Tyrannulus elatus*		Un	R	LC
Greenish Elaenia	*Myiopagis viridicata*		Un	R	LC
Yellow-bellied Elaenia	*Elaenia flavogaster*		Fc	R	LC

English Name	Scientific Name	A	B	C	D
Lesser Elaenia	*Elaenia chiriquensis*		Fc	R	LC
Mountain Elaenia	*Elaenia frantzii*		Fc	R	LC
Yellow-bellied Tyrannulet	*Ornithion semiflavum*		R	R	LC
Brown-capped Tyrannulet	*Ornithion brunneicapillus*		Un	R	LC
Northern Beardless Tyrannulet	*Camptostoma imberbe*		Fc	R	LC
Southern Beardless Tyrannulet	*Camptostoma obsoletum*		Fc	R	LC
Mouse-coloured Tyrannulet	*Phaeomyias murina*		R	R?	LC
Torrent Tyrannulet	*Serpophaga cinerea*		Fc	R	LC
Yellow Tyrannulet	*Capsiempis flaveola*		C	R	LC
Cocos Flycatcher	*Nesotriccus ridgwayi*	Ec	C	Rc	VU
Mistletoe Tyrannulet	*Zimmerius parvus*		C	R	LC
Rufous-browed Tyrannulet	*Phylloscartes superciliaris*		R	R	LC
Olive-striped Flycatcher	*Mionectes olivaceus*		Fc	R	LC
Ochre-bellied Flycatcher	*Mionectes oleagineus*		Fc	R	LC
Sepia-capped Flycatcher	*Leptopogon amaurocephalus*		R	R	LC
Slaty-capped Flycatcher	*Leptopogon superciliaris*		Fc	R	LC
Northern Scrub Flycatcher	*Sublegatus arenarum*		Un	R	LC
Bran-coloured Flycatcher	*Myiophobus fasciatus*		Fc	R	LC
Black-capped Pygmy Tyrant	*Myiornis atricapillus*		Fc	R	LC
Northern Bentbill	*Oncostoma cinereigulare*		Fc	R	LC
Scale-crested Pygmy Tyrant	*Lophotriccus pileatus*		C	R	LC
Slaty-headed Tody-flycatcher	*Poecilotriccus sylvia*		Fc	R	LC
Common Tody-flycatcher	*Todirostrum cinereum*		C	R	LC
Black-headed Tody-flycatcher	*Todirostrum nigriceps*		Un	R	LC
Eye-ringed Flatbill	*Rhynchocyclus brevirostris*		Un	R	LC
Yellow-olive Flatbill	*Tolmomyias sulphurescens*		C	R	LC
Yellow-margined Flatbill	*Tolmomyias flavotectus*		Un	R	LC
Stub-tailed Spadebill	*Platyrinchus cancrominus*		Un	R	LC
White-throated Spadebill	*Platyrinchus mystaceus*		Fc	R	LC
Golden-crowned Spadebill	*Platyrinchus coronatus*		C	R	LC
Tawny-chested Flycatcher	*Aphanotriccus capitalis*		Un	R	VU
Black Phoebe	*Sayornis nigricans*		C	R	LC
Eastern Phoebe	*Sayornis phoebe*		R	V	LC
Vermilion Flycatcher	*Pyrocephalus rubinus*		R	V	LC
Northern Tufted Flycatcher	*Mitrephanes phaeocercus*		C	R	LC
Olive-sided Flycatcher	*Contopus cooperi*		Un	M	NT
Dark Pewee	*Contopus lugubris*	Ep	Un	R	LC
Ochraceous Pewee	*Contopus ochraceus*	Ep	R	R	LC
Western Wood Pewee	*Contopus sordidulus*		C	M	LC
Eastern Wood Pewee	*Contopus virens*		C	M	LC
Tropical Pewee	*Contopus cinereus*		C	R	LC
Yellow-bellied Flycatcher	*Empidonax flaviventris*		Fc	M	LC
Acadian Flycatcher	*Empidonax virescens*		Fc	M	LC
Willow Flycatcher	*Empidonax traillii*		C	M	LC
Alder Flycatcher	*Empidonax alnorum*		C	M	LC
White-throated Flycatcher	*Empidonax albigularis*		Un	R	LC
Least Flycatcher	*Empidonax minimus*		R	M	LC
Yellowish Flycatcher	*Empidonax flavescens*		Fc	R	LC
Black-capped Flycatcher	*Empidonax atriceps*	Ep	Fc	R	LC
Long-tailed Tyrant	*Colonia colonus*		Fc	R	LC
Piratic Flycatcher	*Legatus leucophaius*		C	M	LC
Social Flycatcher	*Myiozetetes similis*		C	R	LC
Rusty-margined Flycatcher	*Myiozetetes cayanensis*		R	R	LC
Grey-capped Flycatcher	*Myiozetetes granadensis*		C	R	LC
Great Kiskadee	*Pitangus sulphuratus*		C	R	LC
White-ringed Flycatcher	*Conopias albovittatus*		Fc	R	LC
Golden-bellied Flycatcher	*Myiodynastes hemichrysus*	Ep	Un	R	LC
Sulphur-bellied Flycatcher	*Myiodynastes luteiventris*		Fc	M	LC

English Name	Scientific Name	A	B	C	D
Streaked Flycatcher	*Myiodynastes maculatus*		Fc	R	LC
Boat-billed Flycatcher	*Megarynchus pitangua*		Fc	R	LC
Tropical Kingbird	*Tyrannus melancholicus*		C	R	LC
Western Kingbird	*Tyrannus verticalis*		Un	M	LC
Scissor-tailed Flycatcher	*Tyrannus forficatus*		C	M	LC
Fork-tailed Flycatcher	*Tyrannus savana*		Fc	R	LC
Eastern Kingbird	*Tyrannus tyrannus*		C	M	LC
Grey Kingbird	*Tyrannus dominicensis*		R	M	LC
Rufous Mourner	*Rhytipterna holerythra*		Fc	R	LC
Dusky-capped Flycatcher	*Myiarchus tuberculifer*		C	R	LC
Panamanian Flycatcher	*Myiarchus panamensis*		Un	R	LC
Ash-throated Flycatcher	*Myiarchus cin*[illegible]		R	V	LC
Nutting's Flycatcher	*Myiarchus nuttingi*		Un	R	LC
Great Crested Flycatcher	*Myiarchus crinitus*		C	M	LC
Brown-crested Flycatcher	*Myiarchus tyrannulus*		C	R	LC
Bright-rumped Attila	*Attila spadiceus*		C	R	LC
Cotingas	**Cotingidae**				
Lovely Cotinga	*Cotinga amabilis*		R	R	LC
Turquoise Cotinga	*Cotinga ridgwayi*	Ep	Un	R	VU
Three-wattled Bellbird	*Procnias tricarunculatus*		Fc	R	VU
Rufous Piha	*Lipaugus unirufus*		Fc	R	LC
Snowy Cotinga	*Carpodectes nitidus*		Un	R	LC
Yellow-billed Cotinga	*Carpodectes antoniae*	Ep	Un	R	EN
Purple-throated Fruitcrow	*Querula purpurata*		Fc	R	LC
Bare-necked Umbrellabird	*Cephalopterus glabricollis*	Ep	Un	R	EN
Manakins	**Pipridae**				
White-ruffed Manakin	*Corapipo altera*		C	R	LC
Blue-crowned Manakin	*Lepidothrix coronata*		C	R	LC
White-collared Manakin	*Manacus candei*		C	R	LC
Orange-collared Manakin	*Manacus aurantiacus*	Ep	Fc	R	LC
Long-tailed Manakin	*Chiroxiphia linearis*		Fc	R	LC
Lance-tailed Manakin	*Chiroxiphia lanceolata*		Un	R	LC
White-crowned Manakin	*Dixiphia pipra*		Un	R	LC
Red-capped Manakin	*Dixiphia mentalis*		C	R	LC
Tityras and Becards	**Tityridae**				
Sharpbill	Oxyruncus cristatus		R	R	LC
Northern Royal Flycatcher	*Onychorhynchus mexicanus*		Un	R	LC
Sulphur-rumped Myiobius	*Myiobius sulphureipygius*		Fc	R	LC
Black-tailed Myiobius	*Myiobius atricaudus*		Un	R	LC
Ruddy-tailed Flycatcher	*Terenotriccus erythrurus*		Un	R	LC
Black-crowned Tityra	*Tityra inquisitor*		Un	R	LC
Masked Tityra	*Tityra semifasciata*		C	R	LC
Northern Schiffornis	*Schiffornis veraepacis*		Un	R	LC
Speckled Mourner	*Laniocera rufescens*		R	R	LC
Barred Becard	*Pachyramphus versicolor*		Fc	R	LC
Cinnamon Becard	*Pachyramphus cinnamomeus*		Fc	R	LC
White-winged Becard	*Pachyramphus polychopterus*		Un	R	LC
Black-and-white Becard	*Pachyramphus albogriseus*		Un	R	LC
Rose-throated Becard	*Pachyramphus aglaiae*		Fc	R	LC
Vireos and Greenlets	**Vireonidae**				
Rufous-browed Peppershrike	*Cyclarhis gujanensis*		Fc	R	LC
Green Shrike-vireo	*Vireolanius pulchellus*		Un	R	LC
White-eyed Vireo	*Vireo griseus*		R	V	LC
Mangrove Vireo	*Vireo pallens*		Un	R	LC
Yellow-throated Vireo	*Vireo flavifrons*		C	M	LC
Blue-headed Vireo	*Vireo solitarius*		R	M	LC
Yellow-winged Vireo	*Vireo carmioli*	Ep	Fc	R	LC
Warbling Vireo	*Vireo gilvus*		Un	M	LC

English Name	Scientific Name	A	B	C	D
Brown-capped Vireo	*Vireo leucophrys*		Fc	R	LC
Philadelphia Vireo	*Vireo philadelphicus*		Fc	M	LC
Red-eyed Vireo	*Vireo olivaceus*		C	M	LC
Yellow-green Vireo	*Vireo flavoviridis*		C	M	LC
Black-whiskered Vireo	*Vireo altiloquus*		R	M	LC
Scrub Greenlet	*Hylophilus flavipes*		Un	R	LC
Tawny-crowned Greenlet	*Hylophilus ochraceiceps*		Fc	R	LC
Lesser Greenlet	*Hylophilus decurtatus*		C	R	LC
Crows and Jays	**Corvidae**				
Azure-hooded Jay	*Cyanolyca cucullata*		Un	R	LC
Silvery-throated Jay	*Cyanolyca argentigula*	Ep	R	R	LC
Black-chested Jay	*Cyanocorax affinis*		Un	R	LC
Brown Jay	*Psilorhinus morio*		C	R	LC
White-throated Magpie-jay	*Calocitta formosa*		C	R	LC
Waxwings	**Bombycillidae**				
Cedar Waxwing	*Bombycilla cedrorum*		Un	M	LC
Silky-flycatchers	**Ptiliogonatidae**				
Black-and-yellow Phainoptila	*Phainoptila melanoxantha*	Ep	Fc	R	LC
Long-tailed Silky-flycatcher	*Ptiliogonys caudatus*	Ep	Fc	R	LC
Swallows and Martins	**Hirundinidae**				
Sand Martin (Bank Swallow)	*Riparia riparia*		C	M	LC
Tree Swallow	*Tachycineta bicolor*		R	M	LC
Mangrove Swallow	*Tachycineta albilinea*		C	R	LC
Violet-green Swallow	*Tachycineta thalassina*		R	V	LC
Purple Martin	*Progne subis*		Un	M	LC
Grey-breasted Martin	*Progne chalybea*		Fc	R	LC
Brown-chested Martin	*Progne tapera*		R	V	LC
Blue-and-white Swallow	*Notiochelidon cyanoleuca*		C	R	LC
Northern Rough-winged Swallow	*Stelgidopteryx serripennis*		Fc	R	LC
Southern Rough-winged Swallow	*Stelgidopteryx ruficollis*		Fc	R	LC
Barn Swallow	*Hirundo rustica*		C	M	LC
American Cliff Swallow	*Petrochelidon pyrrhonota*		C	M	LC
Cave Swallow	*Petrochelidon fulva*		R	V	LC
Wrens	**Troglodytidae**				
Band-backed Wren	*Campylorhynchus zonatus*		Fc	R	LC
Rufous-backed Wren	*Campylorhynchus capistratus*		C	R	LC
Rock Wren	*Salpinctes obsoletus*		R	R	LC
Sedge Wren	*Cistothorus platensis*		Un	R	LC
Black-throated Wren	*Pheugopedius atrogularis*		Fc	R	LC
Black-bellied Wren	*Pheugopedius fasciatoventris*		Fc	R	LC
Spot-breasted Wren	*Pheugopedius maculipectus*		Un	R	LC
Rufous-breasted Wren	*Pheugopedius rutilus*		Fc	R	LC
Banded Wren	*Thryophilus pleurostictus*		Fc	R	LC
Rufous-and-white Wren	*Thryophilus rufalbus*		C	R	LC
Cabanis's Wren	*Cantorchilus modestus*		C	R	LC
Canebrake Wren	*Cantorchilus zeledoni*		Fc	R	LC
Isthmian Wren	*Cantorchilus elutus*		Fc	R	LC
Riverside Wren	*Cantorchilus semibadius*	Ep	C	R	LC
Bay Wren	*Cantorchilus nigricapillus*		C	R	LC
Stripe-breasted Wren	*Cantorchilus thoracicus*		C	R	LC
House Wren	*Troglodytes aedon*		C	R	LC
Ochraceous Wren	*Troglodytes ochraceus*	Ep	Fc	R	LC
Timberline Wren	*Thryorchilus browni*	Ep	Fc	R	LC
White-breasted Wood Wren	*Henicorhina leucosticta*		C	R	LC
Grey-breasted Wood Wren	*Henicorhina leucophrys*		C	R	LC
Northern Nightingale-Wren	*Microcerculus philomela*		Fc	R	LC
Southern Nightingale-wren	*Microcerculus marginatus*		Fc	R	LC
Song Wren	*Cyphorhinus phaeocephalus*		Un	R	LC

English Name	Scientific Name	A	B	C	D
Gnatcatchers	**Polioptilidae**				
Tawny-faced Gnatwren	*Microbates cinereiventris*		Fc	R	LC
Long-billed Gnatwren	*Ramphocaenus melanurus*		C	R	LC
White-lored Gnatcatcher	*Polioptila albiloris*		Un	R	LC
Tropical Gnatcatcher	*Polioptila plumbea*		C	R	LC
Mockingbirds and Thrashers	**Mimidae**				
Grey Catbird	*Dumetella carolinensis*		Fc	M	LC
Tropical Mockingbird	*Mimus gilvus*		Un	R	LC
Thrushes	**Turdidae**				
Black-faced Solitaire	*Myadestes melanops*	Ep	C	R	LC
Black-billed Nightingale-thrush	*Catharus gracilirostris*	Ep	C	R	LC
Orange-billed Nightingale-thrush	*Catharus aurantiirostris*		Fc	R	LC
Slaty-backed Nightingale-thrush	*Catharus fuscater*		Fc	R	LC
Ruddy-capped Nightingale-thrush	*Catharus frantzii*		Fc	R	LC
Black-headed Nightingale-thrush	*Catharus mexicanus*		Fc	R	LC
Veery	*Catharus fuscescens*		Un	M	LC
Grey-cheeked Thrush	*Catharus minimus*		Un	M	LC
Swainson's Thrush	*Catharus ustulatus*		C	M	LC
Wood Thrush	*Hylocichla mustelina*		C	M	NT
Sooty Thrush	*Turdus nigrescens*	Ep	C	R	LC
Mountain Thrush	*Turdus plebejus*		Fc	R	LC
Pale-vented Thrush	*Turdus obsoletus*		Fc	R	LC
Clay-coloured Thrush	*Turdus grayi*		C	R	LC
White-throated Thrush	*Turdus assimilis*		Fc	R	LC
Dippers	**Cinclidae**				
American Dipper	*Cinclus mexicanus*		Fc	R	LC
Old World Sparrows	**Passeridae**				
House Sparrow	*Passer domesticus*		C	I	LC
Tricolored Munia	*Lonchura malacca*		Un	I	LC
Wagtails and Pipits	**Motacillidae**				
Buff-bellied Pipit	*Anthus rubescens*		R	V	LC
Finches	**Fringillidae**				
Lesser Goldfinch	*Spinus psaltria*		Un	R	LC
Yellow-bellied Siskin	*Spinus xanthogastra*		Un	R	LC
Scrub Euphonia	*Euphonia affinis*		Fc	R	LC
Yellow-crowned Euphonia	*Euphonia luteicapilla*		C	R	LC
Thick-billed Euphonia	*Euphonia laniirostris*		Un	R	LC
Yellow-throated Euphonia	*Euphonia hirundinacea*		C	R	LC
Elegant Euphonia	*Euphonia elegantissima*		Un	R	LC
Spot-crowned Euphonia	*Euphonia imitans*	Ep	Un	R	LC
Olive-backed Euphonia	*Euphonia gouldi*		C	R	LC
White-vented Euphonia	*Euphonia minuta*		Un	R	LC
Tawny-capped Euphonia	*Euphonia anneae*		Fc	R	LC
Golden-browed Chlorophonia	*Chlorophonia callophrys*	Ep	Fc	R	LC
New World Warblers	**Parulidae**				
Ovenbird	*Seiurus aurocapilla*		Fc	M	LC
Worm-eating Warbler	*Helmitheros vermivorum*		Un	M	LC
Louisiana Waterthrush	*Parkesia motacilla*		Un	M	LC
Northern Waterthrush	*Parkesia noveboracensis*		C	M	LC
Golden-winged Warbler	*Vermivora chrysoptera*		Fc	M	NT
Blue-winged Warbler	*Vermivora cyanoptera*		R	M	LC
Black-and-white Warbler	*Mniotilta varia*		C	M	LC
Prothonotary Warbler	*Protonotaria citrea*		Fc	M	LC
Flame-throated Warbler	*Oreothlypis gutturalis*	Ep	Fc	R	LC
Tennessee Warbler	*Leiothlypis peregrina*		C	M	LC
Orange-crowned Warbler	*Leiothlypis celata*		R	V	LC
Nashville Warbler	*Leiothlypis ruficapilla*		R	M	LC
Connecticut Warbler	*Oporornis agilis*		R	V	LC

English Name	Scientific Name	A	B	C	D
Grey-crowned Yellowthroat	*Geothlypis poliocephala*		C	R	LC
Chiriqui Yellowthroat	*Geothlypis chiriquensis*		R	R	LC
MacGillivray's Warbler	*Geothlypis tolmiei*		Un	M	LC
Mourning Warbler	*Geothlypis philadelphia*		Fc	M	LC
Kentucky Warbler	*Geothlypis formosa*		Un	M	LC
Olive-crowned Yellowthroat	*Geothlypis semiflava*		Fc	R	LC
Common Yellowthroat	*Geothlypis trichas*		Un	M	LC
Hooded Warbler	*Setophaga citrina*		Un	M	LC
American Redstart	*Setophaga ruticilla*		Fc	M	LC
Slate-throated Whitestart	*Myioborus miniatus*		C	R	LC
Collared Whitestart	*Myioborus torquatus*	Ep	C	R	LC
Cape May Warbler	*Setophaga tigrina*		R	M	LC
Cerulean Warbler	*Setophaga cerulea*		Un	M	VU
Northern Parula	*Setophaga americana*		R	M	LC
Tropical Parula	*Setophaga pitiayumi*		Fc	R	LC
Magnolia Warbler	*Setophaga magnolia*		Un	M	LC
Bay-breasted Warbler	*Setophaga castanea*		Fc	M	LC
Blackburnian Warbler	*Setophaga fusca*		C	M	LC
American Yellow Warbler	*Setophaga aestiva*		C	M	LC
Mangrove Warbler	*Setophaga petechia*		C	R	LC
Chestnut-sided Warbler	*Setophaga pensylvanica*		C	R	LC
Blackpoll Warbler	*Setophaga striata*		R	M	LC
Black-throated Blue Warbler	*Setophaga caerulescens*		R	M	LC
Palm Warbler	*Setophaga palmarum*		R	M	LC
Pine Warbler	*Setophaga pinus*		R	M	LC
Myrtle Warbler	*Setophaga coronata*		Fc	M	LC
Audubon's Warbler	*Setophaga auduboni*		Un	M	LC
Yellow-throated Warbler	*Setophaga dominica*		R	M	LC
Prairie Warbler	*Setophaga discolor*		R	M	LC
Townsend's Warbler	*Setophaga townsendi*		R	M	LC
Hermit Warbler	*Setophaga occidentalis*		R	M	LC
Golden-cheeked Warbler	*Setophago chrysoparia*		R	V	EN
Black-throated Green Warbler	*Setophaga virens*		C	M	LC
Buff-rumped Warbler	*Myiothlypis fulvicauda*		C	R	LC
Rufous-capped Warbler	*Basileuterus rufifrons*		C	R	LC
Black-cheeked Warbler	*Basileuterus melanogenys*	Ep	Fc	R	LC
Golden-crowned Warbler	*Basileuterus culicivorus*		Fc	R	LC
Three-striped Warbler	*Basileuterus tristriatus*		C	R	LC
Canada Warbler	*Cardellina canadensis*		C	M	LC
Wilson's Warbler	*Cardellina pusilla*		C	M	LC
Family uncertain	**Incertae Sedis 2**				
Wrenthrush	*Zeledonia coronata*	Ep	Fc	R	LC
Yellow-breasted Chat	*Icteria virens*		R	M	LC
Oropendolas, Orioles and Blackbirds	**Icteridae**				
Chestnut-headed Oropendola	*Psarocolius wagleri*		Un	R	LC
Crested Oropendola	*Psarocolius decumanus*		R	R	LC
Montezuma Oropendola	*Psarocolius montezuma*		C	R	LC
Scarlet-rumped Cacique	*Cacicus microrhynchus*		C	R	LC
Yellow-billed Cacique	*Amblycercus holosericeus*		Fc	R	LC
Spot-breasted Oriole	*Icterus pectoralis*		Un	R	LC
Yellow-tailed Oriole	*Icterus mesomelas*		Fc	R	LC
Black-cowled Oriole	*Icterus prosthemelas*		Fc	R	LC
Orchard Oriole	*Icterus spurius*		C	M	LC
Baltimore Oriole	*Icterus galbula*		C	M	LC
Bullock's Oriole	*Icterus bullockii*		Un	M	LC
Streak-backed Oriole	*Icterus pustulatus*		Fc	R	LC
Giant Cowbird	*Molothrus oryzivorus*		Fc	R	LC
Bronzed Cowbird	*Molothrus aeneus*		Fc	R	LC

English Name	Scientific Name	A	B	C	D
Shiny Cowbird	*Molothrus bonariensis*		Un	R	LC
Melodious Blackbird	*Dives dives*		Fc	R	LC
Red-winged Blackbird	*Agelaius phoeniceus*		C	R	LC
Great-tailed Grackle	*Quiscalus mexicanus*		C	R	LC
Nicaraguan Grackle	*Quiscalus nicaraguensis*		Un	R	LC
Red-breasted Blackbird	*Sturnella militaris*		Fc	R	LC
Eastern Meadowlark	*Sturnella magna*		Fc	R	LC
Yellow-headed Blackbird	*Xanthocephalus xanthocephalus*		R	V	LC
Bobolink	*Dolichonyx oryzivorus*		R	M	LC
Bananaquit	**Coerebidae**				
Bananaquit	*Coereba flaveola*		C	R	LC
Buntings and New World Sparrows	**Emberizidae**				
Lincoln's Sparrow	*Melospiza lincolnii*		R	V	LC
Rufous-collared Sparrow	*Zonotrichia capensis*		C	R	LC
Volcano Junco	*Junco vulcani*	Ep	Fc	R	LC
Savannah Sparrow	*Passerculus sandwichensis*		R	V	LC
Grasshopper Sparrow	*Ammodramus savannarum*		Un	R	LC
Chipping Sparrow	*Spizella passerina*		R	V	LC
Clay-coloured Sparrow	*Spizella pallida*		R	V	LC
Lark Sparrow	*Chondestes grammacus*		R	V	LC
Stripe-headed Sparrow	*Peucaea ruficauda*		C	R	LC
Botteri's Sparrow	*Peucaea botterii*		R	R	LC
Rusty Sparrow	*Aimophila rufescens*		R	R	LC
Cabanis's Ground Sparrow	*Melozone cabanisi*		Un	R	LC
White-eared Ground Sparrow	*Melozone leucotis*		Fc	R	LC
Olive Sparrow	*Arremonops rufivirgatus*		C	R	LC
Black-striped Sparrow	*Arremonops conirostris*		C	R	LC
Orange-billed Sparrow	*Arremon aurantiirostris*		C	R	LC
Chestnut-capped Brush Finch	*Arremon brunneinucha*		Fc	R	LC
Costa Rican Brush Finch	*Arremon costaricensis*	Ep	Un	r	LC
Sooty-faced Finch	*Arremon crassirostris*	Ep	Un	R	LC
Large-footed Finch	*Pezopetes capitalis*	Ep	Fc	R	LC
White-naped Brush Finch	*Atlapetes albinucha*		Fc	R	LC
Yellow-thighed Finch	*Pselliophorus tibialis*	Ep	Fc	R	LC
Common Bush Tanager	*Chlorospingus flavopectus*		C	R	LC
Sooty-capped Bush Tanager	*Chlorospingus pileatus*	Ep	Fc	R	LC
Ashy-throated Bush Tanager	*Chlorospingus canigularis*		Un	R	LC
Tanagers and allies	**Thraupidae**				
Dusky-faced Tanager	*Mitrospingus cassinii*		Fc	R	LC
Grey-headed Tanager	*Eucometis penicillata*		Fc	R	LC
White-shouldered Tanager	*Tachyphonus luctuosus*		Fc	R	LC
Tawny-crested Tanager	*Tachyphonus delatrii*		Fc	R	LC
White-lined Tanager	*Tachyphonus rufus*		Fc	R	LC
White-throated Shrike-tanager	*Lanio leucothorax*		Un	R	LC
Crimson-collared Tanager	*Ramphocelus sanguinolentus*		Fc	R	LC
Passerini's Tanager	*Ramphocelus passerinii*		C	R	LC
Cherrie's Tanager	*Ramphocelus costaricensis*	E	C	R	LC
Blue-grey Tanager	*Thraupis episcopus*		C	R	LC
Yellow-winged Tanager	*Thraupis abbas*		R	V	LC
Palm Tanager	*Thraupis palmarum*		C	R	LC
Blue-and-gold Tanager	*Bangsia arcaei*	Ep	Un	R	NT
Plain-coloured Tanager	*Tangara inornata*		Un	R	LC
Emerald Tanager	*Tangara florida*		Un	R	LC
Silver-throated Tanager	*Tangara icterocephala*		C	R	LC
Speckled Tanager	*Tangara guttata*		Fc	R	LC
Bay-headed Tanager	*Tangara gyrola*		C	R	LC
Rufous-winged Tanager	*Tangara lavinia*		Un	R	LC
Golden-hooded Tanager	*Tangara larvata*		C	R	LC

English Name	Scientific Name	A	B	C	D
Spangle-cheeked Tanager	*Tangara dowii*	Ep	Fc	R	LC
Scarlet-thighed Dacnis	*Dacnis venusta*		Fc	R	LC
Blue Dacnis	*Dacnis cayana*		Fc	R	LC
Shining Honeycreeper	*Cyanerpes lucidus*		Fc	R	LC
Red-legged Honeycreeper	*Cyanerpes cyaneus*		C	R	LC
Green Honeycreeper	*Chlorophanes spiza*		Fc	R	LC
Sulphur-rumped Tanager	*Heterospingus rubrifrons*	Ep	R	R	LC
Black-and-yellow Tanager	*Chrysothlypis chrysomelas*	Ep	Fc	R	LC
Slaty Flowerpiercer	*Diglossa plumbea*	Ep	C	R	LC
Slaty Finch	*Haplospiza rustica*		R	R	LC
Peg-billed Finch	*Acanthidops bairdi*		Un	R	LC
Grassland Yellow Finch	*Sicalis luteola*		R	V	LC
Wedge-tailed Grass Finch	*Emberizoides herbicola*		R	R	LC
Blue-black Grassquit	*Volatinia jacarina*		C	R	LC
Slate-coloured Seedeater	*Sporophila schistacea*		R	R	LC
Variable Seedeater	*Sporophila corvina*		C	R	LC
White-collared Seedeater	*Sporophila torqueola*		Fc	R	LC
Yellow-bellied Seedeater	*Sporophila nigricollis*		Un	R	LC
Ruddy-breasted Seedeater	*Sporophila minuta*		Un	R	LC
Thick-billed Seed Finch	*Oryzoborus funereus*		Un	R	LC
Nicaraguan Seed Finch	*Oryzoborus nuttingi*		Un	R	LC
Yellow-faced Grassquit	*Tiaris olivaceus*		C	R	LC
Cocos Finch	*Pinaroloxias inornata*	Ec	C	Rc	VU
Rosy Thrush-tanager	*Rhodinocichla rosea*		Un	R	LC
Cardinals, Grosbeaks and allies	**Cardinalidae**				
Flame-coloured Tanager	*Piranga bidentata*		Fc	R	LC
Tooth-billed Tanager	*Piranga lutea*		Un	R	LC
Summer Tanager	*Piranga rubra*		C	M	LC
Scarlet Tanager	*Piranga olivacea*		Fc	M	LC
Western Tanager	*Piranga ludoviciana*		Un	M	LC
White-winged Tanager	*Piranga leucoptera*		Un	R	LC
Red-crowned Ant Tanager	*Habia rubica*		Un	R	LC
Red-throated Ant Tanager	*Habia fuscicauda*		Fc	R	LC
Black-cheeked Ant Tanager	*Habia atrimaxillaris*	E	Fc	R	EN
Carmiol's Tanager	*Chlorothraupis carmioli*		Fc	R	LC
Dickcissel	*Spiza americana*		Fc	M	LC
Black-thighed Grosbeak	*Pheucticus tibialis*	Ep	Fc	R	LC
Rose-breasted Grosbeak	*Pheucticus ludovicianus*		Fc	M	LC
Black-headed Grosbeak	*Pheucticus melanocephalus*		R	V	LC
Black-faced Grosbeak	*Caryothraustes poliogaster*		C	R	LC
Blue Seedeater	*Amaurospiza concolor*		R	R	LC
Slate-coloured Grosbeak	*Saltator grossus*		Un	R	LC
Black-headed Saltator	*Saltator atriceps*		C	R	LC
Buff-throated Saltator	*Saltator maximus*		C	R	LC
Greyish Saltator	*Saltator coerulescens*		Fc	R	LC
Streaked Saltator	*Saltator striatipectus*		Fc	R	LC
Blue-black Grosbeak	*Cyanocompsa cyanoides*		Fc	R	LC
Blue Grosbeak	*Passerina caerulea*		R	M	LC
Indigo Bunting	*Passerina cyanea*		Un	M	LC
Painted Bunting	*Passerina ciris*		Un	M	NT

WEBSITES AND CONTACTS

The following websites should be of interest to anyone contemplating a visit to Costa Rica.

BIRDWATCHING AND TRAVELLING WITHIN COSTA RICA

Wildlife Lodges
www.wildlife-lodges.com
The world's number one website for finding accommodation related to wildlife anywhere in the world. Search by species of bird, mammal, reptile, butterfly, amphibian, dragonfly, plant or sea life, or by country, hotspot or map to find the perfect place to stay to see a particular species.

Costa Rica Gateway
www.costaricagateway.com
Provides an excellent service for birders and nature lovers from around the world. Run by the country's leading birders, it offers invaluable information to help set up your perfect birdwatching and wildlife holiday.

Costa Rica Expeditions
www.costaricaexpeditions.com
Travel agent and tour operator that offers a wide variety of services, including birdwatching and nature tours led by the most respected and experienced guide in the country – Charlie Gomez.

Cotinga Tours
www.cotingatours.com
Highly rated birdwatching and nature tour company with the super Paco Madrigal as master guide. His experience and knowledge of the country's wildlife and superb spotting skills have made him one of Costa Rica's finest guides.

OTHER USEFUL ONLINE RESOURCES

Xeno-Canto
www.xeno-canto.org
Vast online collection of bird sounds covering most of the world's birds, and a good starting point to familiarize yourself with the bird calls of Costa Rica and other countries.

Neotropical Birds
http://neotropical.birds.cornell.edu/portal/home
Superb website with lots of information on bird species found throughout Central and South America. Photographs, maps, sounds and detailed information make this an invaluable resource.

eBird
www.ebird.org
Superb global online platform for birdwatchers to share sight records and checklists.

The IOC World List
www.worldbirdnames.org
This book follows the latest version of Master List (v. 9.1) and conforms to both English and scientific names. Note the use of the English spelling 'grey' rather than the American 'gray' throughout.

SUGGESTED READING

Beletsky, Les, *Costa Rica: Ecotravellers' Wildlife Guide*, 1998, Academic Press Inc.
Chacon, Federico Munez & Richard Dennis Johnston, *Amphibians and Reptiles of Costa Rica*, 2013, Comstock Publishing Associates.
Garrigues, Richard & Robert Dean, *The Birds of Costa Rica*, 2014 (2nd edn), Helm Field Guides.
Henderson, Carrol L., *Mammals, Amphibians and Reptiles of Costa Rica*, 2011, University of Texas Press.
Lawson, Barrett, *Where to Watch Birds in Costa Rica*, 2010, Christopher Helm Publishers Ltd.
Wainwright, Mark, *The Mammals of Costa Rica*, 2007, Cornell University Press.

ACKNOWLEDGEMENTS

Thanks must go to those who helped in the production of this book: Gina Nichol, Frank W. Mantlik, John Schwarz, Jim Carr and John Ashworth for additional photos, and Harry Bird for help with the global status and current up-to-date checklist.